Ben Stacy Jerrik (Ed.)

Innertkirchen

Ben Stacy Jerrik (Ed.)

Innertkirchen

Communes of Switzerland, Communes of Switzerland, Cantons of Switzerland

Part Press

Imprint

Publisher:
Part Press is a trademark of
International Book Market Service Ltd., 17 Rue Meldrum, Beau Bassin, 1713-01 Mauritius
Email: info@bookmarketservice.com
Website: www.bookmarketservice.com

Published in 2013

Printed in: U.S.A., U.K., Germany. This book was not produced in Mauritius.

ISBN: 978-620-1-59337-4

Contents

Articles

Innertkirchen 1

Communes_of_Switzerland 4

Interlaken-Oberhasli_(administrative_district) 9

Cantons_of_Switzerland 11

Canton_of_Bern 18

Coat_of_arms 33

Agricultural_cooperative 40

Kerns 46

Schattenhalb 50

Wolfenschiessen 53

Guttannen 57

Gadmen 60

Grindelwald 63

Engelberg 71

References

Article Sources and Contributors 76

Image Sources, Licenses and Contributors 78

Innertkirchen

Innertkirchen	
 View of Innertkirchen from the river Aar	
Country Switzerland **Canton** Bern **District** Interlaken-Oberhasli 	
Population	837 (Dec 2011)[1]
- Density	7 /km^2 (18 /sq mi)
Area	km^2 (sq mi)[2]
Elevation	625 m (ft)
Postal code	3862
SFOS number	0784
Surrounded by	Engelberg (OW), Gadmen, Grindelwald, Guttannen, Hasliberg, Kerns (OW), Meiringen, Schattenhalb, Wolfenschiessen (NW)
Website	www.innertkirchen.ch [3] SFSO statistics [4]

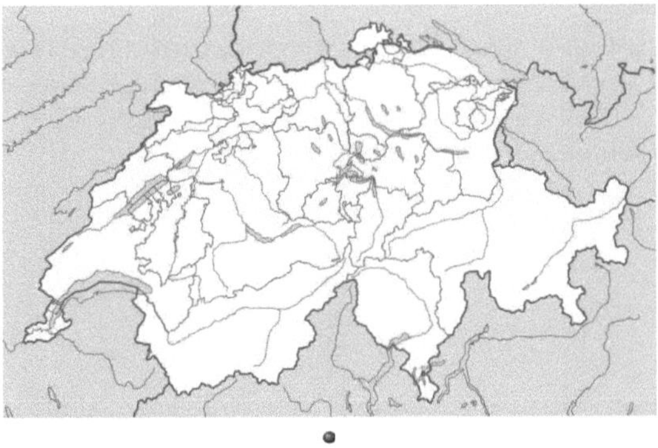

Innertkirchen

Innertkirchen is a village and municipality in the Interlaken-Oberhasli administrative district in the canton of Bern in Switzerland.

Its coat of arms depicts an eagle wearing an imperial crown.

History

The name of the municipality is of recent origin and first appeared in print in 1834.[5] It was formed from the farming settlements (*Bäuerten*) of Grund, Bottigen, Wyler-Schattseite, Wyler-Sonnseite and the agricultural cooperative of Äppigen.

Geography

Innertkirchen has an area, as of 2009, of 120.05 km^2 (sq mi). Of this area, 18.77 km^2 (sq mi) or 15.6% is used for agricultural purposes, while 24.15 km^2 (sq mi) or 20.1% is forested. Of the rest of the land, 0.92 km^2 (sq mi) or 0.8% is settled (buildings or roads), 1.97 km^2 (sq mi) or 1.6% is either rivers or lakes and 74.18 km^2 (sq mi) or 61.8% is unproductive land.[6]

Of the built up area, housing and buildings made up 0.3% and transportation infrastructure made up 0.3%. 13.8% of the total land area is heavily forested. Of the agricultural land, 2.4% is pastures and 13.2% is used for alpine pastures. All the water in the municipality is in rivers and streams. Of the unproductive areas, 11.5% is unproductive vegetation, 29.0% is too rocky for vegetation and 21.3% of the land is covered by glaciers.[6]

Demographics

Innertkirchen has a population (as of 31 December 2011) of 837.[1] As of 2007, 4.4% of the population was made up of foreign nationals. Over the last 10 years the population has decreased at a rate of -6.9%. Most of the population (as of 2000) speaks German (95.4%), with Portuguese being second most common (1.1%) and Italian being third (0.6%).

In the 2007 election the most popular party was the SVP which received 59.6% of the vote. The next three most popular parties were the SPS (11.3%), the FDP (10.3%) and the Green Party (6.7%).

The age distribution of the population (as of 2000) is children and teenagers (0–19 years old) make up 24% of the population, while adults (20–64 years old) make up 56.3% and seniors (over 64 years old) make up 19.7%. The entire Swiss population is generally well educated. In Innertkirchen about 71.2% of the population (between age 25-64) have completed either non-mandatory upper secondary education or additional higher education (either university or a *Fachhochschule*).

Innertkirchen has an unemployment rate of 0.47%. As of 2005, there were 127 people employed in the primary economic sector and about 39 businesses involved in this sector. 252 people are employed in the secondary sector and there are 16 businesses in this sector. 150 people are employed in the tertiary sector, with 35 businesses in this sector.[7] The historical population is given in the following table:[5]

year	population
1764	656
1850	1,375
1880	1,438
1900	1,105
1950	1,194
1960	1,230

Industry

The village has access to considerable reserves of running water and two hydropower plants provide around 1,500 million kWh of power per annum, as well as employment for 220 persons.

Transportation

Innertkirchen is the terminus of the local Meiringen-Innertkirchen Bahn narrow-gauge railway, which links the village with Meiringen railway station in the nearby town of Meiringen. Connection can be made here with the Brünigbahn railway from Interlaken to Lucerne.

Centre of Innertkirchen, at the crossroads between the Susten and Grimsel Pass

References

[1] Swiss Federal Statistical Office - STAT-TAB (http://www.pxweb.bfs.admin.ch/Database/German_01 - BevÃ¶lkerung/01.2 - BevÃ¶lkerungsstand und -bewegung/01.2 - BevÃ¶lkerungsstand und -bewegung.asp?lang=1&prod=01&secprod=2&openChild=true), online database – *Datenwürfel für Thema 01.2 - Bevölkerungsstand und -bewegung* (German) accessed 3 October 2012

[2] Arealstatistik Standard - Gemeindedaten nach 4 Hauptbereichen (http://www.bfs.admin.ch/bfs/portal/de/index/themen/02/03/blank/data/gemeindedaten.html)

[3] http://www.innertkirchen.ch

[4] http://www.bfs.admin.ch/bfs/portal/en/index/regionen/02/key.html

[5] *Innertkirchen* in German (http://www.hls-dhs-dss.ch/textes/d/D470.php), French (http://www.hls-dhs-dss.ch/textes/f/F470.php) and Italian (http://www.hls-dhs-dss.ch/textes/i/I470.php) in the online *Historical Dictionary of Switzerland*.

[6] Swiss Federal Statistical Office-Land Use Statistics (http://www.bfs.admin.ch/bfs/portal/de/index/themen/02/03/blank/data/gemeindedaten.html) 2009 data (German) accessed 25 March 2010

[7] Swiss Federal Statistical Office (http://www.bfs.admin.ch/bfs/portal/en/index/regionen/02/key.html) accessed 10-Jul-2009

External links

- (German) http://www.innertkirchen.ch - official site
- Dubler, A-M: *Innertkirchen* in German (http://www.hls-dhs-dss.ch/textes/d/D470.php), French (http://www.hls-dhs-dss.ch/textes/f/F470.php) and Italian (http://www.hls-dhs-dss.ch/textes/i/I470.php) in the online *Historical Dictionary of Switzerland*, 2006-05-01.

Communes_of_Switzerland

Communes (German: *Gemeinden* / French: *communes* / Italian: *comuni* / Romansh: *vischnancas*), also known as municipalities, are the smallest government division in Switzerland, numbering 2,596 (as of February 2010).[1] While many have a population of a few hundred citizens, the largest cities such as Zürich or Geneva also have the legal status of municipalities. The area of the municipalities varies between 0.28 km² (Ponte Tresa, Ticino) and 430 km² (Glarus Süd, Glarus).

Map of Switzerland showing municipal boundaries (2012)

Each canton defines the responsibilities of its constituent communes. These may include providing local government services such as education, medical and social services, public transportation, and tax collection. The degree of centralization varies from one canton to another.

Communes are generally governed by a council (sometimes called Municipality) headed by a mayor as executive and the town meeting as legislature. Most cantons leave the option to larger municipalities to opt for a city parliament. In some cantons, foreign persons that have dwelled for a certain time in Switzerland are also allowed to participate in the municipal politics.

Swiss citizenship is based on the citizenship of a municipality. Every Swiss is citizen of one or several municipalities (known as the place of origin, *lieu d'origine*, or *Heimatort*). As on the cantonal and federal level, citizens enjoy their political rights, including the direct democratic ones, in their municipality.

Communes are financed through direct taxes (such as the income tax), with rates varying more or less within a framework set by the canton. As among the cantons, there is a tax transfer among the communes to balance various levels of tax income.

Many municipalities are having difficulties maintaining the civil services they need to perform the duties they are required to do. In an effort to reduce expenses, some municipalities are combining together (through mergers or the creation of special-purpose districts). This restructuring is generally encouraged by the cantonal governments and the rate of these unions is increasing.

"Cities" (*villes* or *Städte*) are the municipalities with more than 10'000 inhabitants, or smaller places which had medieval town rights. There is no specific designation for smaller communities such as "village" or "town".

Population	Number of municipalities in 2004 (%)[2]
>20,000	30 (1.1%)
10,000–19,999	89 (3.2%)
5000–9999	180 (6.6%)
1000–4999	1025 (37.4%)
500–999	555 (20.3%)
<500	861 (31.4%)
Total	2740 (100%)

Communes are numbered by the Swiss Federal Office for Statistics (see *Community Identification Number#Switzerland*). One or more postal codes (PLZ/NPA) can by assigned to a municipality or shared with other municipalities.

See also: List of cities in Switzerland

Lists of communes by canton

• Aargau: *Einwohnergemeinden*	• Grisons: *politische Gemeinden, Vischnancas politicas* or *comuni politici*	• St. Gallen: *Politische Gemeinden*
• Appenzell I.R.: *Bezirke*	• Jura: *communes*	• Thurgau: *politische Gemeinden*
• Appenzell A.R.: *Einwohnergemeinden*	• Lucerne: *Einwohnergemeinden*	• Ticino: *comuni politici*
• Basel-City: *Einwohnergemeinden*	• Neuchâtel: *communes*	• Uri: *Einwohnergemeinde*
• Basel-Country: *Einwohnergemeinden*	• Nidwalden: *Gemeinden*	• Valais: *communes municipales* or *Einwohnergemeinden*
• Bern: *Einwohnergemeinden* or *communes municipales*	• Obwalden: *Einwohnergemeinden*	• Vaud: *communes*
• Fribourg: *communes* or *Gemeinden*	• Schaffhausen: *Einwohnergemeinden*	• Zug: *Einwohnergemeinde*
• Geneva: *communes*	• Schwyz: *Gemeinden*	• Zürich: *Politische Gemeinden*
• Glarus: *Ortsgemeinden*	• Solothurn: *Einwohnergemeinden*	

See cantons of Switzerland for the number of municipalities per canton.

List of communes by population

Largest municipalities

City	C.	Pop.
Zürich	ZH	376,815
Geneva	GE	185,726
Basel	BS	167,365
Bern	BE	128,041
Lausanne	VD	129,273
Winterthur	ZH	100,000
St. Gallen	SG	74,538
Lucerne	LU	57,890
Lugano	TI	56,719

Biel/Bienne	BE	50,852
Thun	BE	42,319
Köniz	BE	38,098
La Chaux-de-Fonds	NE	37,023
Schaffhausen	SH	33,527
Fribourg	FR	33,008
Chur	GR	34,915
Neuchâtel	NE	32,389
Vernier	GE	32,038
Uster	ZH	28,770
Sion	VS	27,697
Emmen	LU	27,274

Smallest municipalities (As of 2002)

Rank	Municipality	C.	Pop.
1.	Portein	GR	22
2.	Corippo	TI	23
3.	Mulegns	GR	25
4.	Bister	VS	26
5.	Martisberg	VS	28
6.	Goumoens-le-Jux	VD	30
6.	Largario	TI	30
8.	Clugin	GR	35
9.	Cauco	GR	36
9.	Monible	BE	36
11.	St. Martin	GR	37
12.	Gresso	TI	38
12.	Montfavergier	JU	38
14.	Vaugondry	VD	39
15.	Arrissoules	VD	40
15.	Romairon	VD	40
15.	Selma	GR	40
18.	Ausserbinn	VS	41
18.	Linescio	TI	41
18.	Mauraz	VD	41
18.	Steinhaus	VS	41

nb. Corippo has now taken place as smallest city/municipality in Switzerland with only 17 inhabitants. (2004 Census)

History

The beginnings of the modern municipality system date back to the Helvetic Republic. Under the Old Swiss Confederacy, citizenship was granted by each town and village to only residents. These citizens enjoyed access to community property and in some cases additional protection under the law. Additionally, the urban towns and the rural villages had differing rights and laws. The creation of a uniform Swiss citizenship, which applied equally for citizens of the old towns and their tenants and servants, led to conflict. The wealthier villagers and urban citizens held rights rights to forests, common land and other municipal property which they did not want to share with the "new citizens", who were generally poor. The compromise solution, which was written into the municipal laws of the Helvetic Republic, is still valid today. Two politically separate but often geographically similar organizations were created. The first, the so-called municipality, was a political community formed by election and its voting body consists of all resident citizens. However, the community land and property remained with the former local citizens who were gathered together into the *Bürgergemeinde*. During the Mediation era (1803–1814), and especially during the Restoration era (1814–1830), many of the gains toward uniform citizenship were lost. Many political municipalities were abolished and limits were placed on the exercise of political rights for everyone except the members of the *Bürgergemeinde*. In the Regeneration era (1830–1848), the liberal revolutions of the common people helped to restore some rights again in a few cantons. In other cantons, the *Bürgergemeinden* were able to maintain power as political communities. In the city of Zurich it wasn't until the Municipal Act of 1866 that the political municipality came back into existence.[3]

The relationship between the political municipality and the *Bürgergemeinde* was often dominated by the latter's ownership of community property. Often the administration and profit from the property were totally held by the *Bürgergemeinden*, leaving the political municipality dependent on the *Bürgergemeinde* for money and use of the property. It wasn't until the political municipality acquired rights over property that served the public (such as schools, fire stations, etc.) and taxes, that they obtained full independence. For example, in the city of Bern, it wasn't until after the property division of 1852 that the political municipality had the right to levy taxes.[3]

It wasn't until the Federal Constitution of 1874 that all Swiss citizens were granted equal political rights on local and Federal levels. This revised constitution finally removed all the political voting and electoral body rights from the *Bürgergemeinde*. In the cities, the percentage of members in the *Bürgergemeinde* in the population was reduced as a result of increasing emigration to the cities. This led to the *Bürgergemeinde* losing its former importance to a large extent. However, the *Bürgergemeinde* has remained, and it includes all individuals who are citizens of the *Bürgergemeinde*, usually by having inherited the *Bürgerrecht* (citizenship), regardless of where they were born or where they may currently live. Instead of the place of birth, Swiss legal documents, e.g. passports, contain the *Bürgerort* (place of citizenship). The *Bürgergemeinde* also often holds and administers the common property in the village for the members of the community.[3]

Other types of communes

In addition to the political communes or municipalities, a number of other communes exist in Switzerland. These include:

- Bürgergemeinde (also: Burgergemeinde, Ortsgemeinde, Ortsbürgergemeinde, Tagwen, bourgeoisie, commune bourgeoise, vischnanca burgaisa), a statutory corporation that includes everyone who is a citizen of a commune and has the *Heimatrecht* (home right) in that commune regardless of where they may currently live. Until the 19th Century this *Heimatrecht* included rights to use the commons, which were administered by the Bürgergemeinde. Modernly, some Bürgergemeinden may still control common property, but the *Heimatrecht* and associated *Heimatort* is used just as place of birth in other countries.
- Gemischte Gemeinde (mixed communes), found in the Canton of Jura and portions of the Canton of Bern, a combination of a Bürgergemeinde and a political commune.

- Korporationsgemeinde, a legally recognized cooperative in Central Switzerland that controls some land and is responsible for its members support.
- Kirchgemeinde, a parish for members of a large church (generally Roman Catholic or Swiss Reformed. There may be two or more *Kirchgemeinden* in a single municipality.
- Schulgemeinde, similar to a School district.
- Bäuert, in the Berner Oberland or Graubünden) a small farming community. It is a type of agricultural cooperative with shared equipment and land.[4]
- Degagna, in the Leventina valley in the Canton of Ticino. It manages shared pastures, fields and woods as well as maintaining roads that cross the common land.[5]

See also

- List of municipalities of Switzerland
- List of twin towns and sister cities in Switzerland
- Former municipalities of Switzerland

Notes and references

[2] Official list of Swiss municipalities, p. 17

[3] *Bürgergemeinde* in German (http://www.hls-dhs-dss.ch/textes/d/D26443.php), French (http://www.hls-dhs-dss.ch/textes/f/F26443.php) and Italian (http://www.hls-dhs-dss.ch/textes/i/I26443.php) in the online *Historical Dictionary of Switzerland*.

[4] *Bäuert* in German (http://www.hls-dhs-dss.ch/textes/d/D10405.php), French (http://www.hls-dhs-dss.ch/textes/f/F10405.php) and Italian (http://www.hls-dhs-dss.ch/textes/i/I10405.php) in the online *Historical Dictionary of Switzerland*.

[5] *Degagna* in German (http://www.hls-dhs-dss.ch/textes/d/D35272.php), French (http://www.hls-dhs-dss.ch/textes/f/F35272.php) and Italian (http://www.hls-dhs-dss.ch/textes/i/I35272.php) in the online *Historical Dictionary of Switzerland*.

External links

- Amtliches Gemeindeverzeichnis der Schweiz/Liste officielle des communes de la Suisse/Elenco ufficiale dei Comuni della Svizzera, 2006 (http://www.bfs.admin.ch/bfs/portal/de/index/news/publikationen.html?publicationID=2321). Neuchâtel, 2006. ISBN 3-303-00334-3.
- *Nachbarschaft* in German (http://www.hls-dhs-dss.ch/textes/d/D16403.php), French (http://www.hls-dhs-dss.ch/textes/f/F16403.php) and Italian (http://www.hls-dhs-dss.ch/textes/i/I16403.php) in the online *Historical Dictionary of Switzerland*.

Interlaken-Oberhasli_(administrative_district)

Interlaken-Oberhasli District in the Canton of Bern was created on 1 January 2010, uniting the former Interlaken and Oberhasli districts.[1] It is part of the Oberland administrative region. It contains 29 municipalities with an area of 1228.93 km^2 (sq mi) and a population (as of December 2008) of 48,763.

Flag	Name	Population (31 December 2011)[2]	Area in km²
	Beatenberg	1139	29.18
	Bönigen	2446	15.14
	Brienz	2973	47.94
	Brienzwiler	513	17.62
	Därligen	417	6.89
	Gadmen	228	116.30
	Grindelwald	3796	170.60
	Gsteigwiler	420	7.01
	Gündlischwand	290	16.88
	Guttannen	300	200.77
	Habkern	632	51.06
	Hasliberg	1236	41.68
	Hofstetten bei Brienz	555	8.81
	Innertkirchen	837	120.10
	**Interlaken**	5468	4.31
	Iseltwald	424	22.12
	Lauterbrunnen	2509	164.47
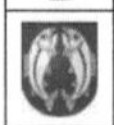	Leissigen	948	10.40
	Lütschental	240	13.21

	Matten bei Interlaken	3757	5.92
	Meiringen	4602	40.17
	Niederried bei Interlaken	345	4.24
	Oberried am Brienzersee	468	20.11
	Ringgenberg	2627	8.81
	Saxeten	101	19.09
	Schattenhalb	577	31.41
	Schwanden bei Brienz	593	7.031
	Unterseen	5505	14.05
	Wilderswil	2535	13.61
	Total (29)	46481	1228.93

References

[1] Amtliches Gemeindeverzeichnis der Schweiz, Mutationsmeldungen 2009 / Répertoire officiel des communes de Suisse, Mutations 2009 / Elenco ufficiale dei Comuni della Svizzera, Mutazione 2009 (http://www.bfs.admin.ch/bfs/portal/de/index/infothek/nomenklaturen/blank/blank/gem_liste/03.Document.128095.pdf) (Report). Federal Statistical Office. 2009. 2776. . Retrieved 6 March 2010.

[2] Swiss Federal Statistical Office - STAT-TAB (http://www.pxweb.bfs.admin.ch/Database/German_01 - BevÃ¶lkerung/01.2 - BevÃ¶lkerungsstand und -bewegung/01.2 - BevÃ¶lkerungsstand und -bewegung.asp?lang=1&prod=01&secprod=2&openChild=true), online database – *Datenwürfel für Thema 01.2 - Bevölkerungsstand und -bewegung* (German) accessed 3 October 2012

This article incorporates information from the equivalent article on the German Wikipedia.

Cantons_of_Switzerland

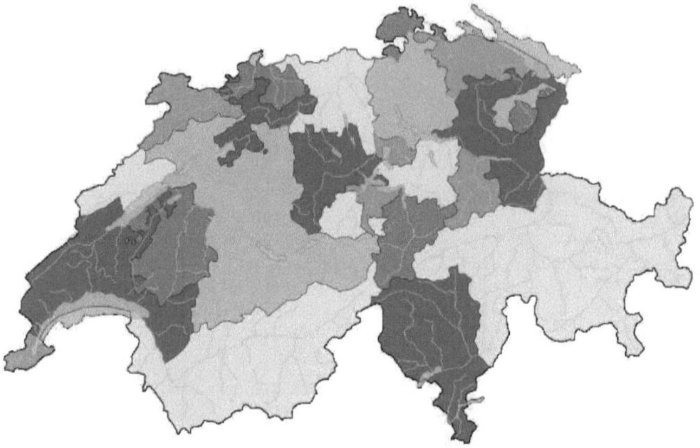

Valais

Ticino

Graubünden (Grisons)

Geneva

Vaud

Neuchâtel

Jura

Bern

Thurgau

Zurich

Aargau

Lucerne

Solothurn

Basel-Land

Schaffhausen

Uri

Schwyz

Glarus

St. Gallen

AI

AR

Obwalden

Nidwalden

Zug

Fribourg

Basel-Stadt

France

Italy

Liecht.

Austria

Germany

The 26 **cantons of Switzerland** are the member states of the federal state of Switzerland. Each canton was a fully sovereign state[1] with its own border controls, army and currency from the Treaty of Westphalia (1648) until the establishment of the Swiss federal state in 1848. The most recently created canton is the Canton of Jura, which separated from the Canton of Bern in 1979.[2]

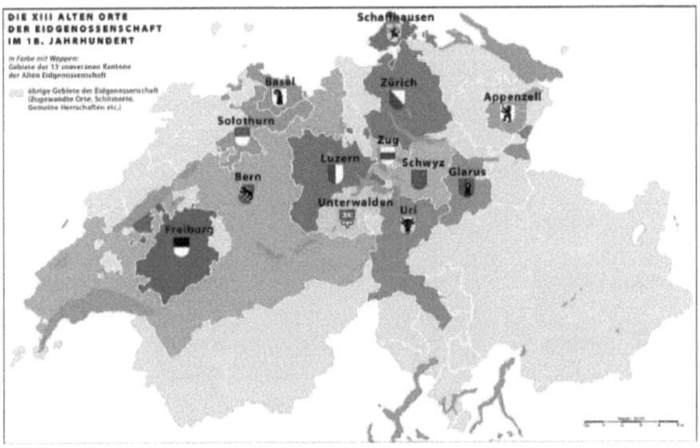

The 13 cantons of the Old Swiss Confederacy (1513-1798).

The name is derived from the French language word *canton* meaning *corner* or *district* (from which the term Cantonment is also derived).

History

In the 16th century, the Old Swiss Confederacy was composed of 13 sovereign cantons, and there were two different kinds: six land (or forest) cantons and seven city (or urban) cantons. Though they were technically part of the Holy Roman Empire, they had become *de facto* independent when the Swiss defeated Emperor Maximillian in 1499.[] The six forest cantons were democratic republics, whereas the seven urban cantons were oligarchic republics controlled by noble families.

Constitution

Each canton has its own constitution, legislature, government and courts.[3] Most of the cantons' legislatures are unicameral parliaments, their size varying between 58 and 200 seats. A few legislatures are general assemblies known as *Landsgemeinden*. The cantonal governments consist of either five or seven members, depending on the canton.[4] For the names of the institutions, see List of legislative and executive councils of the Cantons of Switzerland.

The Swiss Federal Constitution declares the cantons to be sovereign to the extent their sovereignty is not limited by federal law.[3] The cantons also retain all powers and competencies not delegated to the Confederation by the Constitution. Most significantly, the cantons are responsible for healthcare, welfare, law enforcement and public education; they also retain the power of taxation. The cantonal constitutions determine the degree of autonomy accorded to the municipalities, which varies but almost always includes the power to levy taxes and pass municipal laws. The sizes of the cantons vary from 37 km² to 7,105 km²; the populations vary from 15,471 to 1,244,400.

Direct democracy

As on the federal level, all cantons provide for (half-) direct democracy. Citizens may demand a popular vote to amend the cantonal constitution or laws, or to veto laws or spending bills passed by the parliament. General popular assemblies (*Landsgemeinde*) are now limited to the cantons of Appenzell Innerrhoden and Glarus. In all other cantons democratic rights are exercised by secret ballot.

List

The cantons are listed in the order given in the federal constitution.[a]

Coat of arms	Abbr	Canton	Since	Capital	Population[1]	Area (km²) [2]	Density (per km²) [3]	No. munic.[4]	Official languages
	ZH	**Zurich**	1351	Zurich	1406083		701	171	German
	BE	**Bern**	1353	Bern	985046		158	383	German, French
	LU	**Lucerne**	1332	Lucerne	381966		233	87	German
	UR	**Uri**	1291[5]	Altdorf	35382		33	20	German
	SZ	**Schwyz**	1291[5]	Schwyz	147904		143	30	German
	OW	**Obwalden**	1291[5] or 1315 (as part of Unterwalden)	Sarnen	35878		66	7	German
	NW	**Nidwalden**	1291[5] (as Unterwalden)	Stans	41311		138	11	German
	GL	**Glarus**	1352	Glarus	39217		51	3	German
	ZG	**Zug**	1352	Zug	116559		416	11	German
	FR	**Fribourg**	1481	Fribourg	284668		141	167	French, German
	SO	**Solothurn**	1481	Solothurn	259836		308	122	German
	BS	**Basel-Stadt**	1501 (as Basel until 1833/1999)	Basel	194090		5,072	3	German
	BL	**Basel-Landschaft**	1501/1833[6]	Liestal	277973		502	86	German
	SH	**Schaffhausen**	1501	Schaffhausen	77139		246	27	German
	AR	**Appenzell Ausserrhoden**	1513 [7]	Herisau[5]	53313		220	20	German
	AI	**Appenzell Innerrhoden**	1513[7]	Appenzell	15789		87	6	German
	SG	**St. Gallen**	1803[8]	St. Gallen	483101		222	85	German
	GR	**Graubünden**	1803[9]	Chur	193388		26	180	German, Romansh, Italian
	AG	**Aargau**	1803	Aarau	627893		388	220	German

	TG	**Thurgau**	1803[10]	Frauenfeld6	254528		229	80	German
	TI	**Ticino**	1803[11]	Bellinzona	336943		110	157	Italian
	VD	**Vaud**	1803[12]	Lausanne	725944		188	339	French
	VS	**Valais**	1815[13]	Sion	317022		53	143	French, German
	NE	**Neuchâtel**	1815/1857[14]	Neuchâtel	173183		206	53	French
	GE	**Geneva**	1815	Geneva	472530		1,442	45	French
	JU	**Jura**	1979[15]	Delémont	70542		82	64	French
	CH	**Switzerland**		Bern	7,968,705		174	2,596	German, French, Italian, Romansh

The two-letter abbreviations for Swiss cantons are widely used, e.g. on car license plates. They are also used in the ISO 3166-2 codes of Switzerland with the prefix "CH-" (*Confœderatio Helvetica*, Switzerland), e.g. CH-SZ for the canton of Schwyz.

Half-cantons

Six of the 26 cantons are traditionally, but no longer officially, called "half-cantons" (German: *Halbkanton*, French: *demi-canton*, Italian: *semicantone*), reflecting a history of mutual association or partition.

The half-cantons are identified in the first article of the Swiss Federal Constitution of 1999 by being joined to their other "half" with the conjunction "and":

> The People and the Cantons of Zurich, Bern, Lucerne, Uri, Schwyz, Obwalden and Nidwalden, Glarus, Zug, Fribourg, Solothurn, Basel-Stadt and Basel-Landschaft, Schaffhausen, Appenzell Ausserrhoden and Appenzell Innerrhoden, St. Gallen, Graubünden, Aargau, Thurgau, Ticino, Vaud, Valais, Neuchâtel, Geneva, and Jura form the Swiss Confederation.
>
> —Article 1 of the Federal Constitution of the Swiss Confederation (underlining not in original)[16]

The 1999 constitutional revision retained this distinction, on the request of the six cantonal governments, as a way to mark the historic association of the half-cantons to each other.[17] In contrast, the first article of the 1848 and 1874 constitutions constituted the Confederation as the union of "twenty-two sovereign cantons",[18] referring to the half-cantons as "Unterwalden (above and beneath the woods)", "Basel (city and country)" and "Appenzell (both Rhoden)".[19] While the older constitutions referred to these states as "half-cantons", a term that remains in popular use, the 1999 revision and official terminology since then use the appellation "cantons with half a cantonal vote".[20]

With their mutual association a purely historical matter, the half-cantons are since 1848 equal to the other cantons in all but two respects:[21]

- They elect only one member of the Council of States instead of two (Cst. art. 150 par. 2).
- In popular referendums about constitutional amendments, which require for adoption a national popular majority as well as the assent of a majority of the cantons (*Ständemehr / majorité des cantons*), the result of the half-cantons' popular vote counts only one half of that of the other cantons (Cst. arts. 140, 142). This means that for purposes of a constitutional referendum, at least 12 out of a total of 23 cantonal popular votes must support the amendment.[22]

The reasons for the association between the three pairs of half-cantons are varied:

Caricature of the division of Basel, 1833

- *Unterwalden* never consisted of a single unified jurisdiction. Originally, Obwalden, Nidwalden, and the Abbey of Engelberg formed distinct communities. The collective term *Unterwalden* remains in use, however, for the area that partook in the creation of the original Swiss confederation in 1291 with Uri and Schwyz. The Federal Charter of 1291 called for representatives from each of the three "areas".[23] [24]
- The canton of Appenzell divided itself into an "inner" and "outer" half ("Rhoden") as a consequence of the Reformation in Switzerland in 1597:[25] Appenzell Innerrhoden (Catholic) and Appenzell Ausserrhoden (Protestant).
- The canton of Basel divided itself as a consequence of a revolt of the Basel countryside in 1833, in order to promote equality among its citizenry, combating claims between rural and city residents over preferential status:[26] Basel-Landschaft and Basel-Stadt.

Names in national languages

(Names appear in bold when corresponding to the cantonal official language)

Abbr	Common English	Other English forms	German	French	Italian	Romansh
AG	Aargau	Argovia	**Aargau**	Argovie	Argovia	Argovia
AI	Appenzell Innerrhoden	Appenzell Inner-Rhodes	**Appenzell Innerrhoden**	Appenzell Rhodes-Intérieures	Appenzello Interno	Appenzell dadens
AR	Appenzell Ausserrhoden	Appenzell Outer-Rhodes	**Appenzell Ausserrhoden**	Appenzell Rhodes-Extérieures	Appenzello Esterno	Appenzell dador
BS	Basel-Stadt	Basle-City	**Basel-Stadt**	Bâle-Ville	Basilea-Città	Basilea-Citad
BL	Basel-Landschaft	Basle-Country	**Basel-Landschaft**	Bâle-Campagne	Basilea-Campagna	Basilea-Champagna
BE	Bern	Berne	**Bern**	**Berne**	Berna	Berna
FR	Fribourg	Friburg	**Freiburg**	**Fribourg**	Friborgo	Friburg
GE	Geneva	-	Genf	**Genève**	Ginevra	Genevra
GL	Glarus	Glaris	**Glarus**	Glaris	Glarona	Glaruna
GR	Graubünden	Grisons	**Graubünden**	Grisons	**Grigioni**	**Grischun**
JU	Jura	-	Jura	**Jura**	Giura	Giura
LU	Lucerne	-	**Luzern**	Lucerne	Lucerna	Lucerna
NE	Neuchâtel	-	Neuenburg	**Neuchâtel**	Neuchâtel	Neuchâtel
NW	Nidwalden	Nidwald	**Nidwalden**	Nidwald	Nidvaldo	Sutsilvania
OW	Obwalden	Obwald	**Obwalden**	Obwald	Obvaldo	Sursilvania
SH	Schaffhausen	Schaffhouse	**Schaffhausen**	Schaffhouse	Sciaffusa	Schaffusa
SZ	Schwyz	-	**Schwyz**	Schwyz (or Schwytz)	Svitto	Sviz
SO	Solothurn	Soleure	**Solothurn**	Soleure	Soletta	Soloturn
SG	St. Gallen	St. Gall	**St. Gallen**	Saint-Gall	San Gallo	Son Gagl

TG	Thurgau	Thurgovia	**Thurgau**	Thurgovie	Turgovia	Turgovia
TI	Ticino	Tessin	Tessin	Tessin	**Ticino**	Tessin
UR	Uri	-	**Uri**	Uri	Uri	Uri
VS	Valais	Wallis	**Wallis**	**Valais**	Vallese	Vallais
VD	Vaud	-	Waadt	**Vaud**	Vaud	Vad
ZG	Zug	-	**Zug**	Zoug	Zugo	Zug
ZH	Zurich	-	**Zürich**	Zurich	Zurigo	Turitg

Admission of new cantons

The enlargement of Switzerland by way of the admission of new cantons ended in 1815. After a failed attempt of Vorarlberg to join Switzerland in 1919, the idea of resuming Swiss enlargement was revived in 2010 by a parliamentary motion that would allow the accession of regions bordering on Switzerland.

See also

- List of legislative and executive councils of the Cantons of Switzerland
- Data codes for Switzerland#Cantons
- List of cantons of Switzerland by elevation
- Flags of Swiss cantons
- Cantonal bank, a commercial bank (at least partially) owned by the canton

Notes

1. This is the order generally used in Swiss official documents. At the head of the list are the three city cantons that were considered preeminent in the Old Swiss Confederacy; the other cantons are listed in order of accession to the Confederation. This traditional order of precedence among the cantons has no practical relevance in the modern federal state, in which the cantons are equal to one another, although it still determines formal precedence among the cantons' officials (see Swiss order of precedence).
2. as of 5 April 2009
3. km^2
4. Per km^2, based on 2000 population
5. As of 31 December 2007,
6. Seat of government and parliament is Herisau, the seat of the judicial authorities is Trogen
7. Seat of parliament half-yearly alternates between Frauenfeld and Weinfelden

References

- (German). Cited as *Ehrenzeller.*
- Cited as *Häfelin.*

[1] *Cantons, In the Old Confederation until 1798* in German (http://www.hls-dhs-dss.ch/textes/d/D26414-1-1.php), French (http://www.hls-dhs-dss.ch/textes/f/F26414-1-1.php) and Italian (http://www.hls-dhs-dss.ch/textes/i/I26414-1-1.php) in the online *Historical Dictionary of Switzerland.*

[2] *Jura (Canton)* in German (http://www.hls-dhs-dss.ch/textes/d/D7399.php), French (http://www.hls-dhs-dss.ch/textes/f/F7399.php) and Italian (http://www.hls-dhs-dss.ch/textes/i/I7399.php) in the online *Historical Dictionary of Switzerland.*

[3] *Cantons, In the Federal State since 1848* in German (http://www.hls-dhs-dss.ch/textes/d/D26414-1-4.php), French (http://www.hls-dhs-dss.ch/textes/f/F26414-1-4.php) and Italian (http://www.hls-dhs-dss.ch/textes/i/I26414-1-4.php) in the online *Historical Dictionary of Switzerland.*

[4] Swiss Government website (http://www.ch.ch/karte/index.html?lang=en) with links to each cantonal government, accessed 11 November 2008

[5] founding forest-canton, foundation date traditionally given as either 1307, 1304 or 1291 (see Foundation of the Old Swiss Confederacy).

[6] part of Basel until 1833/1999

[7] part of Appenzell until 1597/1999

[8] Act of Mediation, formed out of the Canton of Säntis and the northern half of the Canton of Linth.

[9] Act of Mediation; formerly the Canton of Raetia, comprising the earlier Three Leagues.

[10] coterminous with the canton of Thurgau of the Helvetic Republic (1798), formerly a condominium.

[11] combining the former cantons of Bellinzona and Lugano; see *Ennetbirgische Vogteien.*

[12] Act of Mediation, formerly Canton of Léman.

[13] Restoration, formerly the Simplon *département*

[14] claimed by Frederick William III of Prussia until the Neuchâtel Crisis of 1856–1857.

[15] seceded from Berne

[16] Federal Constitution of the Swiss Confederation of 18 April 1999, SR/RS 101 (E (http://www.admin.ch/ch/e/rs/c101.html), D (http://www.admin.ch/ch/d/sr/c101.html), F (http://www.admin.ch/ch/f/rs/c101.html), I (http://www.admin.ch/ch/i/rs/c101.html)), art. 1 (E (http://www.admin.ch/ch/e/rs/101/a1.html), D (http://www.admin.ch/ch/d/sr/101/a1.html), F (http://www.admin.ch/ch/f/rs/101/a1.html), I (http://www.admin.ch/ch/i/rs/101/a1.html))

[17] Felix Hafner / Rainer J. Schweizer in *Ehrenzeller*, Art. 1 N 2; Häfelin, N 966.

[18] Twenty-three after the creation of the Canton of Jura in 1978.

[19] Bundesverfassung der Schweizerischen Eidgenossenschaft vom 29. Mai 1874 (http://www.verfassungen.de/ch/verf74-i.htm), Bundesverfassung der Schweizerischen Eidgenossenschaft vom 12. September 1848 (http://www.verfassungen.de/ch/verf48.htm) (German); author's translation.

[20] Felix Hafner / Rainer J. Schweizer in *Ehrenzeller*, Art. 1 N 10; Häfelin, N 963

[21] Häfelin, N 963, 967

[22] Häfelin, N 950

[23] Pacte fédéral du 1er (http://www.admin.ch/org/polit/00056/index.html?lang=fr) août 1291] sur Admin.ch "vallée inférieure d'Unterwald" signifie Nidwald.

[24] Pacte fédéral du 1er août 1291 (http://hypo.ge.ch/www/cliotexte//html/suisse.histoire.1291.html) sur Cliotexte

[25] Réforme catholique, Contre-Réforme et scission (http://hls-dhs-dss.ch/textes/f/F7389-1-10.php) Article du dictionnaire historique de la Suisse

[26] De la République helvétique à la division du canton (1798-1833) (http://hls-dhs-dss.ch/textes/f/F7387-2-2.php) Article du dictionnaire historique de la Suisse

External links

- Swissworld.org (http://www.swissworld.org/en/politics/general_information/the_cantons/) – The cantons of Switzerland
- GeoPuzzle (http://www.geomit.ch/GeoPuzzles/SchweizPuzzle/schweiz.html) – Assemble cantons on a Swiss map
- Badac (http://www.badac.ch) – Database on Swiss cantons and cities (French/German)

Canton_of_Bern

The **Canton of Bern** (German: *Kanton Bern*, pronunciation; French: *Canton de Berne*) is the second largest of the 26 Swiss cantons by both surface area and population. Located in west-central Switzerland, it borders the Canton of Jura and the Canton of Solothurn to the north. To the west lie the Canton of Neuchâtel, the Canton of Fribourg and Vaud. To the south lies the Valais. East of the canton of Bern lie the cantons of Uri, Nidwalden, Obwalden, Lucerne and Aargau.

The canton of Bern is bilingual and has a population (as of 31 December 2011) of 985046.[1] As of 2007, the population included 119,930 (or 12.45%) foreigners.[2] The cantonal capital, also the federal capital of Switzerland, is Bern.

History

Bern joined the Swiss Confederation in 1353 and was between 1803 and 1814 one of the six *direcorial cantons* of the Napoleonic Swiss Confederation.

Early Prehistory

The earliest traces a human presence in the area of the modern Canton is found in three caves in the Simmental region; Schnurenloch near Oberwil, Ranggiloch above Boltigen and Chilchlihöhle above Erlenbach. These caves were used at various times during the last ice age. The first open-air settlement in the area is an upper paleolithic settlement at Moosbühl in Moosseedorf. During the warmer climate of the mesolithic period, increasing forest cover restricted the movement of hunters, fishers and gatherers. Their temporary settlements were built along lake and marsh edges, which remained free of trees due to fluctuations in water level. Important mesolithic sites in the Canton are at Pieterlenmoos and Burgäschisee lake along with alpine valleys at Diemtig and Simmental. During the neolithic period, there were a number of settlements on the shores of Lake Biel, the Toteisbecken (Lobsigensee, Moossee, Burgäschisee and Inkwilersee) and along rivers (Aare, Zihl).[3] Several of these sites are part of the Prehistoric Pile dwellings around the Alps, a UNESCO World Heritage Site.[4]

One of the best explored neolithic sites is at Twann (now Twann-Tüscherz). In the Twannbach delta there were about 25 Cortaillod culture and Horgen culture villages that existed between 3800 and 2950 BC. One of the oldest examples of bread from Switzerland, a sourdough from 3560–3530 BC, came from one of these villages.[3]

Simple copper objects were already in use in the 4th millennium BC, including a copper pin from Lattrigen from 3170 BC and a knife blade from Twann. Shortly before 2000 BC bronze production entered the area and brought about a surge in development. Settlements began to spread into the pre-Alpine and Alpine areas. The area between Lake Thun and the Niedersimmental were densely settled. Archeological finds include scattered items along mountain passes, a fortified hilltop settlements at Spiezberg, Cholis Grind by Saanen and at Pintel by Wimmis, along with cemeteries at Thun-Allmendingen, Einigen and Hilterfingen. Late Bronze Age (ca. 1000–800 BC) settlements along Lake Biel (Mörigen, Vinelz) have yielded up a wealth of items.

Iron Age

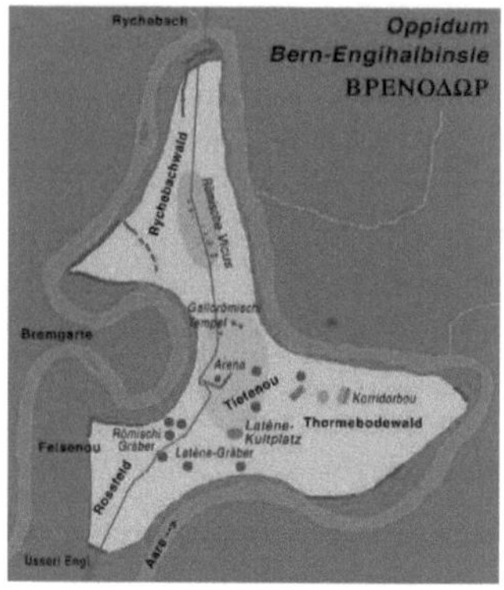

Helveto-roman settlement Bern-Engehalbinsel

During the early Iron Age (Hallstatt culture from 800–450 BC) changes in climate forced them to abandon settlements along many waterways and in the valley floors and move to the plateaus and hills. With increased trade contacts across the Alps, the cultural influence of the Mediterranean grew in the area. Evidence of this trade include a hydria which was discovered in Grächwil. Burial rituals and social classes became more developed during this time. The so-called *princely graves* became more common, many of the burial mounds were over 30 m (ft) in diameter and 4 m (ft) high and richly outfitted with grave goods. In a grave mound in Bützberg the first burial in the mound was followed by several later burials. Often, several grave mounds combined to become a necropolis, such as at Grossaffoltern, Ins, Bannwil, Langenthal and Bützberg. Most of the knowledge about the Hallstatt culture in the Canton comes from graves. The only discovered settlement is around Blanche Church in La Neuveville.

The grave good show that iron was forged into swords, daggers, spearheads, knives and wagon accessories. Gold, which was probably collected from river sand, was made into diadems, rings and pendants. Thin bronze arm, leg and neck plates with geometric designs were often buried, especially in the graves at Allenlüften in Mühleberg, at Ins and at Bützberg. The jewelry that was buried included bracelets and rings which were also made of jet and lignite coal. At Münchringen, the grave pottery was both shaped by hand or thrown on a potter's wheel, and was painted with multi-colored ornamentation.

The transition to the late Iron Age or La Tène period (450–1st century BC) is indicated by a sudden change of style in the metalworking and ceramic industries. Numerous graves (from unknown settlements) along with the two oppida at Bern-Engehalbinsel and Jensberg by Studen mark the population centers during the late Iron Age. Gold coins (from Melchnau) along with silver and bronze coins first start to appear in this same era. A sword with Greek characters that said *Korisios* was found at the Port site. At the oppidum at Bern-Engehalbinsel, there were studios for glass and ceramic production and iron working achieved a high level of skill, along with craftsmen who worked in wood, leather and goldsmithing. There was a nearby place of worship in the Bremgarten wood, and cemeteries at Münsingen and Bern-Engehalbinsel.[3]

Roman era

Baths at Engehalbinsel near Bern

After the Roman victory at Bibracte in 58 BC, the Helvetii were forced to return to their homes as foederati of the Romans. Under increasing Roman influence, the local economy and trade flourished. The main settlements lay on the foot of the Jura Mountains and the Central Plateau. The existing roads were expanded, especially the Aventicum-Vindonissa and the Petinesca-Augusta Raurica roads. A fourth alpine pass, the Rawil was added to the traditional three; the Grimsel, Brünig and Susten passes. In the Bernese Jura the Mont Raimeux and Pierre Pertuis passes opened.

Under the Romans, many of the old fortified places were expanded and refortified. The old Helvetii oppidum at the Engehalbinsel became a Roman vicus, which was probably known as Brenodor or Brenodurum. At the foot of the Jens mountain, the fort Petinesca was built to guard the roads over the Jura Mountains. This was reinforced in the late-roman era (368–369 AD) by a fortified bridge over the Thielle/Zihl river between Aegerten and Brügg.

A number of Roman villas were built around the Canton. At Oberwichtrach both the main building (pars urbana) and the industrial section (pars rustica) of a Roman villa have been discovered. The villas at Münsingen, Toffen and Herzogenbuchsee have rich mosaics that are still partly visible.

The religious practices of the local population merged with Roman beliefs and the Canton of Bern is home to a number of unique cult centers. They include the larger than life statues of gods (including enthroned Jupiter) at Petinesca, Engehalbinsel and Thun-Allmendingen and a number of inscriptions. At the beginning of the 5th century AD, Rome withdrew its troops from the Rhine garrisons, but allowed the Burgundians to settle on Helvetii lands in 443 AD.[3]

Early Middle Ages

Burgundian and Allamanni lands between 534 and 843

East of the Aare river, the transition from the Gallo-Roman dominated population to a Germanic population in the Early Middle Ages happened relatively peacefully. One exception was the Battle of Wangen in 610, but elsewhere it generally was a slow process of cultural infiltration. By the 7th century, the Alamannic settlers had already taken most of the good locations southeast of the Aare and they began moving up the Aare to the regions of Lake Thun and Lake Brienz. All areas west of the Aare river belonged to the romanised Burgundian kingdom, which became part of the Merovingian Frankish kingdom in the 534. During the Carolingian era, the political structure of the Franks had spread into parts of what became Switzerland. In 762/778 the County of Aargau was founded, followed in 861 by the County of Oberaargau and in 965 the County of Bargen. The Treaty of Verdun in 843 put the border between Central and Eastern Frankish Empires at the Aare, and divided the Aare region in half. The population west of the Aare generally spoke a Romance language, while those to the east spoke a Germanic language. The region between the Saane and Aare rivers became the language border.

Christianity spread slowly into the Aare valley. The dioceses of Lausanne, Basel, Sion, Chur and Constance were all established before the new faith made inroads into the valley. The Aare valley was bordered by three dioceses; Lausanne, Constance and Basel. However, the first Christian missionaries came into the valley from Alsace and other western areas. In 630 the Abbey of Luxeuil established the Abbey of Moutier-Grandval along the old transit route through the Pierre Pertuis Pass. By the 9th or 10th Century, this Abbey had property and influence all the way to Lake Biel and into the Balsthal valley. The first monastic cells of what would become the Abbey of Saint-Imier was also founded in 600. In the Seeland and Aare valleys, wooden churches were first built during the Merovingian period. The current churches in Kirchlindach, Oberwil bei Buren and Bleibach were all built above the ruins of these early churches. About 30 churches in the Bern and Solothurn portions of the Aare valley were created over ruins of Roman villas and subsequent burial grounds in the 7th Century (including Meikirch and Oberbipp). In Mett, the church was built over a 5th-century mausoleum, which was built over a 4th-century tomb. In 700, six sarcophagi were buried on St. Peter's Island next to a Roman temple complex. A wooden monastery was built over the complex in the 8th–9th century.[3]

Middle Ages

During the 10th century, the Aare valley came fully under the Second Kingdom of Burgundy. Under King Rudolph I and his son Rudolph II Burgundy's influence reached across most of modern Switzerland. They established royal courts at Bümpliz, Münsingen, Uetendorf, Wimmis, Kirchberg and Utzenstorf to allow them to govern the Aare valley. Later, the Aare valley moved toward closer ties with the Holy Roman Empire during the Ottonian and Salian dynasties.

Lands held by the main noble families around 1200

The succession dispute following the death of Rudolph III in 1032 allowed the Salian kings to acquire the Kingdom of Upper Burgundy and with it the Aare valley. As a part of the Holy Roman Empire, the valley was involved when the Investiture Controversy broke out in 1056. In 1077, the Regent of Burgundy, Count Rudolf of Rheinfelden, declared himself as an anti-king against King Henry IV. The Bishops of Basel and Lausanne remained loyal to King Henry IV, and took the Rheinfelden lands in Oberaargau and the upper Aare valley. After Rudolf's death in 1090, his lands went to his son, Berchtold II of Zähringen. He and his son, Berchtold III, tried to use these lands to expand their power. In 1127 the Zähringens were appointed Rector or delegate of the king in Burgundy. Their hopes for a new, independent Burgundy were dashed in 1156, and the last Zähringen count, Berchtold V, embarked on a program of city founding. The cities of Burgdorf, Murten, Thun and Bern were all founded by Berchtold V. When he died without an heir, the Zähringen lands went to the House of Kyburg, while the offices and fiefs reverted to the empire.[3]

During the High Middle Ages both the Aare valley and the Bernese Oberland were divided into a number of small counties, each with their own baron. Nobles from the Holy Roman Empire began to marry into the local noble families and a number of the Zähringen Ministerialis families (unfree knights in the service of a feudal overlord) moved into the Oberland. The nobles also began to found monasteries to spread their power into the Oberland. During the period between 1070 and 1150, at least ten large monasteries were founded by local nobles. These include; the Cluniac Priory of Rüeggisberg (Lords of Rümligen in 1072), Münchenwiler (Wiler family in 1080), St. Peter's Island (Count of Hochburgund-Mâcon in the late 11th century), Hettiswil (1107), Röthenbach im Emmental (Lords of Rümligen or Signau), the Benedictine monastery at St. Johannsen in Erlach (Fenis family in 1100), Trub (Lords of Lützelflüh before 1130) and Rüegsau (possibly also the Lords of Lützelflüh in first half of the 12th century), the Augustinian Collegiate church in Interlaken (Oberhofen family in 1130) and the Cistercian Frienisberg Abbey (Count Saugern around 1130). However, in 1191 the Oberland barons revolted against Berchtold V of Zähringen and many of the Oberland barons were killed in the battle of Grindelwald.

During the 13th Century a number of the cities near Bern were granted the city right and appointed mayors and city councils. Bern became an imperial city. During the mid-13th century, the Empire's presence weakened in the Aare valley, forcing the local nobles to find allies to protect themselves. Bern entered into a number of treaties with its neighbors in the 13th century. In 1274, the Emperor Rudolph I of Habsburg, confirmed Bern's imperial immediacy. However, in 1285 he imposed an imperial tax which drove the city to support Rudolph's enemies. Although it withstood two sieges by the Emperor in 1288, after the defeat at Schosshalde in 1289 it had to pay taxes and a penalty.

In 1298 Bernese forces won a victory at Oberwangen in Köniz against the County of Savoy and the Habsburg Austrian nobility. In 1300, the city acquired the four surrounding parishes of Bolligen, Vechigen, Stettlen and Muri, destroyed the threatening castles of Bremgarten and Belp and gave the Baron of Montenach Bernese citizenship. After the victory of Louis IV of Bavaria over the Habsburg Frederick the Fair in the battle of Mühldorf (Bavaria) in 1322, Bern entered an alliance with the anti-Habsburg Swiss Forest Cantons in 1323.

Old Swiss Confederacy

The area of the canton of Bern consists of lands acquired by the city of Bern mostly between the 14th and the 16th century, both by conquest and purchase. Acquired districts include (with dates of acquisition):

- Laupen (1324)
- Oberhasli (1334)
- Aarberg (1375)
- Thun and Burgdorf (1384)
- Unterseen and the Upper Simme valley (1386)
- Frutigen and other towns in the Bernese Oberland (1400)
- Aargau (1415)
- Lower Simme valley (1439–1449)
- Aigle and Grandson (1475)
- Interlaken, with Grindelwald, Lauterbrunnen and Brienz (1528, all the suppression of the Augustinian Canons at Interlaken Abbey)
- Vaud (1536)
- Saanen or Gessenay (1555)
- the *Pays d'En-Haut* including Château-d'Œx (1555)
- Köniz (1729).

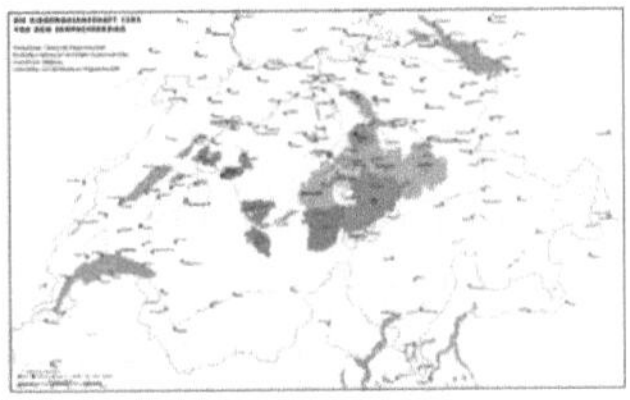

The Swiss Confederacy before the Battle of Sempach (1387)

The Swiss Confederacy in 1416

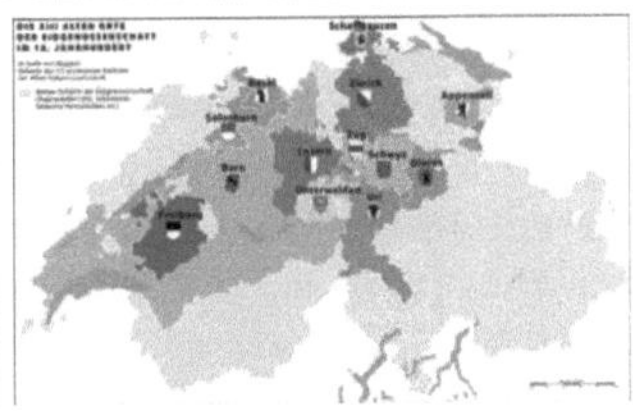

The Swiss Confederacy in the 18th century

The districts of the Bernese Aargau before the creation of the Canton of Aargau

In 1310 Emperor Henry VII pledged Laupen Castle, Laupen and the surrounding lands as collateral for a loan. In 1324, Bern acquired the pledged castle and lands. When the Emperor was unable to repay the loan, Laupen became the first bailiwick of Bern.[5]

In 1322, the brothers Eberhard II of Neu-Kyburg and Hartmann II of Neu-Kyburg started fighting with each other over who would inherit the family's lands around Thun. The fighting led to the "fratricide at Thun Castle" where Eberhard killed his brother Hartmann. To avoid punishment by his Habsburg overlords, Eberhard fled to Bern. In the following year, he sold the town of Thun, its castle and the land surrounding Thun to Bern. Bern then granted the land back to Eberhard as a fief.[6] Bern's support of Eberhard, their resulting expansion into the Oberland and their alliance with the Forest Cantons brought the city into conflict with the Habsburgs during the 14th Century. The Gümmenenkrieg in 1333 between Bern and Fribourg over rights and influence in the Sense/Singine area ended without resolving anything.

In 1334, Bern fought with the Barons of Weissenburg and occupied Wimmis and Unspunnen in the Oberland. Bern's victory allowed them to bring the Oberhasli region, its capital of Meiringen and Weissenburg under their control. Bern's continued expansion was at the expense of the feudal lords in the surrounding lands. In 1339 the Habsburgs, Kyburgs and Fribourg, marched against Bern with 17,000 men and besieged the border town of Laupen. To raise the siege, Bern raised a force of 6,000, consisting of Bernese, supported by the Forest Cantons, and other allies (Simmental, Weissenbur and Oberhasli).[7] The allied Bernese forces were victorious at the Battle of Laupen and Bern drew closer to the Swiss Forest Cantons. It entered into a permanent or eternal alliance with Uri, Schwyz and Unterwalden in 1353. This date is usually considered the date of Bern's entrance into the Swiss Confederation, however the alliance only indirectly tied Bern to Zurich and Lucerne.

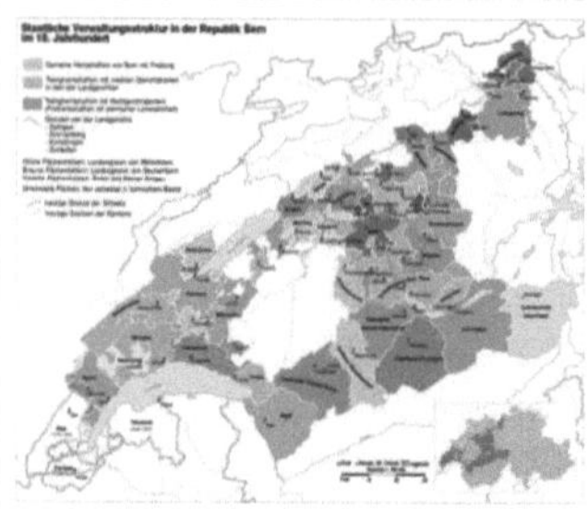

Districts of the Canton of Bern in the 18th Century

By 1400, Bern controlled the entire Bernese Oberland. Under their control, the five valleys of the Oberland enjoyed extensive rights and far-reaching autonomy in the Bäuerten (farming cooperative municipalities) and *Talverbänden* (rural alpine communities). However, throughout the Late Middle Ages, the Oberland, as a whole or in part, revolted several times against Bernese authority. The Evil League (*Böser Bund*) in 1445 fought against Bernese military service and taxes following the Old Zürich War,[8] in 1528 the Oberland rose up in resistance to the Protestant Reformation and in 1641 Thun revolted.[9]

In 1358 the cash-strapped Count Peter II of Aarberg pawned the County of Aarberg to Bern. However, in 1367 he sold it, with out repaying Bern, to his cousin Rudolf IV of Nidau. After Rudolf's death (1375) Bern acquired clear right to the Aarberg lands from the other heirs.[10]

Throughout the 14th Century, Bern entered into alliances and treaties with many of its neighbors, including members of the Swiss Confederation. Bern entered into a series of agreements with Biel/Bienne, which led to a conflict with Biel's ruler, the Prince-Bishop of Basel in 1367–68. The bishop marched south and destroyed Biel along with a number of towns in the southern Jura Mountains. During the Gugler war in 1375, there were several battles between Bernese troops and English mercenaries under Enguerrand de Coucy.

By the 1370s, the Kyburgs (which still held Thun as a fief for Bern) were deeply in debt to Bern. On 11 November 1382, Rudolf II of Neu-Kyburg made an unsuccessful raid on Solothurn. The ensuing conflict with the Old Swiss Confederacy (known as the *Burgdorferkrieg* or *Kyburgerkrieg*) allowed Bern to move against the Habsburgs in Aargau. After the Bernese laid siege to Burgdorf, Neu-Kyburg was forced to concede an unfavourable peace.[11] Bern bought Thun and Burgdorf, the most important cities of Neu-Kyburg, and their remaining towns passed to Bern and Solothurn by 1408. The last of the Neu-Kyburgs, Berchtold, died destitute in Bern in 1417.[6]

In 1386, the Austrians under Leopold of Habsburg invaded eastern Switzerland. When they besieged the city of Sempach, troops from Zurich, Lucerne and the Forest Cantons marched out and defeated the Austrians at the Battle of Sempach. While Bern was not involved at Sempach they used the Austrian weakness to march into the Oberland in 1386, followed by the Seeland (the region south of the Jura Mountains containing the Morat (Murten), Neuchâtel and Bienne (Biel) lakes) in 1388 and the Aargau in 1389. In the peace agreement of 1389 Bern got Unterseen and the Upper Simme valley (Simmental) from the Habsburgs. In the same year, Fribourg accepted Berns acquisition of the Iselgaus, i.e. the area between Lake Biel and the Seeland. Over the next several decades Bern continued to expand to the determent of the Habsburgs and Fribourg. They acquired the following towns: 1391 Simmenegg, 1399 Signau, 1400 Frutigen, 1407 Wangen, 1408 Trachselwald and Huttwil, 1412 Oltigen and in 1413 (together with Solothurn) Bipp and Bechburg.[3]

Siege and execution of the garrison at Grandson

In 1475, during the Burgundian Wars, the mountain regions of Saanen and Pays-d'Enhaut, who were allied with Bern, attacked and burned the tower of Aigle Castle. They then gave Aigle town and the surrounding district to Bern in exchange for not having to pay one-third of their income to Bern. In the treaty of Fribourg from 1476, Fribourg received rights over the Aigle district, which they gave up to Bern in 1483. Bern rebuilt Aigle Castle in 1489 and made it the seat of the bailiwick of Aigle.[12] Also in 1475, the Confederation attacked and captured Grandson Castle. In the next year, Charles the Bold retook the castle and executed the Bernese garrison. In 1476, at the Battle of Grandson, Charles forces retreated and the Bernese retook Grandson. After Charles' defeat at the Battle of Murten, Grandson became a shared territory between Bern and Fribourg. Each city appointed a vogt for five-year periods.[13] Aigle and Grandson were the first French speaking regions in the Canton of Bern.

In the Bernese Oberland during the 13th and the beginning of the 14th century Interlaken Abbey grew to became the largest landholder in the region. The Abbey controlled the towns of Grindlewald, Lauterbrunnen and numerous farms along Lake Brienz. However, in 1350 a period of crises and conflicts led to a decline in the number of monks and nuns and increasing debt. In 1348, the people of Grindelwald and Wilderswil joined a mutual defense league with Unterwalden. Bern responded with a military expedition to the Bernese Oberland, which ended in defeat for Unterwalden and its allies. By 1472, Bern was the patron of the Abbey. During the Protestant Reformation, the Abbey was secularized in 1528. The canons received a financial settlement and the properties were now managed by a Bernese bailiff. The tenants of the Abbey who had expected the abolition of all owed interest, responded by rioting, which was suppressed by Bern.[14]

The lands around the northern shore of Lake Geneva and between Lake Neuchatel were the next area that Bern expanded into. The region had originally been part of the Carolingian Empire. Then, in 1032 the Zähringens of Germany defeated the Burgundians, who were then replaced by the counts of Savoy in 1218. Under the counts of Savoy the area was given political unity, and established as the Barony of Vaud. However, as the power of the Savoys declined at the beginning of the 15th century the land was occupied by troops from Bern. By 1536 the area was completely annexed. Reformation was started by co-workers of John Calvin like Pierre Viret, including a famous debate at the cathedral of Lausanne; but it was only decisively implemented when Bern put its full force behind it. Vaud was another French-speaking region in the mostly German-speaking canton, which caused several uprisings.

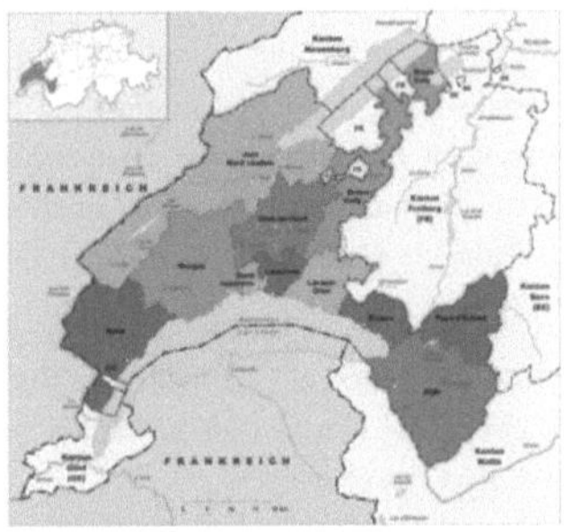

Map of the modern Canton of Vaud, which was annexed by Bern from 1536 until 1798.

Both Château-d'Œx and Saanen were part of the county of Gruyère. Both districts had quite a bit of independence and often entered into treaties against the will and best interests of the Counts. In 1340 the Saanen valley concluded a peace treaty with the Simmental, which provided for arbitration in disputes. They entered into another treaty in 1393 with the Valais. In 1401, Count Rudolph of Gruyère entered into a treaty with Bern which included Saanen. Two years later Saanen[15] and Château-d'Œx[16] negotiated their own alliances with Bern. Due to the Bernese alliance, Saanen sent troops, under their own banner, to support the Bernese invasions of Aargau in 1415 and Valais in 1418. The military losses and taxes following the Old Zürich War led Saanen to support the Evil League (*Böser Bund*) in

1445 against Bern. During the Burgundian War in 1475, Saanen, together with troops from Château-d'Œx and the Simmental captured the Savoy district of Aigle for Bern. Saanen and the surrounding district enjoyed a great deal of independence during the 16th century. However, in 1555 the last Count of Gruyère lost both districts to Bern when his county went bankrupt. Bern took over the entire Saanen valley in the following year and introduced the Protestant Reformation.[15] They incorporated the Pays-d'Enhaut with Château-d'Œx into the new Bernese district of Saanen.[16]

Social changes in early modern Bern

In the Middle Ages, upwards mobility and access to public offices was relatively easy for successful traders and craftsmen, but Bernese society became ever more stratified and aristocratic as the power and wealth of the city grew. By the 17th century, citizenship had become an inherited prerogative, all political bodies elected one another and officials were elected for life.[3] In effect, public offices were now the exclusive prerogative of the *gnädige Herren*, the "merciful lords", as the small number of noble families now ruling Bern came to be called. In 1605 there were 152 families that were qualified to rule, by 1691 that number was only 104, while towards the end of the 18th century there were only 69 such families. Meanwhile the land ruled by the town was extending over more and more territory, so that finally it governed 52 bailiwicks.[] These offices became very lucrative as the Bernese territories grew. Patrician *Landvögte*, sheriffs, ruled the politically powerless countryside, often using armed force to put down peasant revolts.

In Vaud the Bernese occupants were not popular amongst the population. In 1723, Major Abraham Davel led a revolt against Bern, in protest at what he saw as the denial of political rights of the French-speaking Vaudois by the German-speaking Bernese, and was subsequently beheaded.[17]

Napoleonic period

Inspired by the French Revolution, the Vaudois drove out the Bernese governor in 1798 and declared the Lemanic Republic. Vaud nationalists like Frédéric-César de La Harpe had called for French intervention in liberating the area and French Revolutionary troops moved in, taking over the whole of Switzerland itself in the process and setting up the Helvetic Republic.

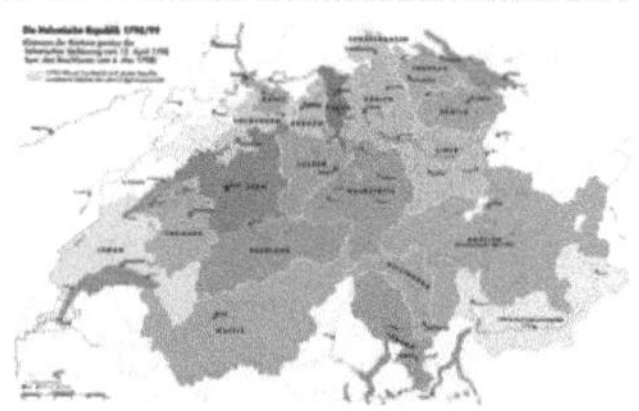

The Helvetic Republic from 1798 to 1801

In 1798, with the establishment of the Helvetic Republic, Bern was divided, the canton of Oberland with Thun as its capital and the canton of Léman with Lausanne as its capital were detached from what was left of the Canton of Bern.

Within the new canton of Oberland, historic borders and traditional rights were not considered. As there had been no previous separatist feeling amongst the conservative population, there was little enthusiasm for the new order.[9] The situation in the canton of Léman was quite different. The French-speaking Vaudois had never felt like part of the German-speaking Canton of Bern. When they joined the Swiss Confederation in 1803, it was as the Canton of Vaud.

Under the Helvetic Republic, Pays-d'Enhaut with Château-d'Œx became part of the Canton of Léman while Saanen and the rest of the district became part of the Canton of Oberland. When the Helvetic Republic collapsed in 1803, Saanen and its district became a district in the new Canton of Bern[15] while Château-d'Œx and its district joined Vaud.

The 1801 Malmaison Constitution proposed reuniting the Oberland with Bern, but it was not until the Act of Mediation, two years later, with the abolition of the Helvetic Republic and the partial restoration of the *ancien régime*, that the two cantons were reunited.[18]

Modern history

Further information: Jurassic separatism

With the Restoration of 1815, Bern acquired the Bernese Jura with Biel/Bienne from the bishopric of Basel, while the canton of Léman became the canton of Vaud and remained separate from Bern.

Bern still remained the largest canton of the confederacy from 1815 to 1979, when parts of the Bernese Jura broke away to form the canton of Jura.

Geography

The canton of Bern is mainly drained by the River Aare and its tributaries. The area of the canton is commonly divided into six regions.[19] The most populated area is the Bernese Mittelland on the plateau north of the Alps, with the capital city of Bern. The northmost part of the canton is the Bernese Jura bordering the Canton of Jura. The Bernese Oberland is the mountainous region which lies in the south of the canton.

The Bernese Alps with Eiger, Mönch and Jungfrau, view from Gurten, Bern in Winter

The area of the canton is 5959 km^2 (sq mi). Of this area, 6.4% is occupied with houses, businesses or roads. 43.3% of the canton is agricultural land, and 31.0% is forested. Less than 19.3% is considered non-productive, which includes glaciers, cliffs and lakes.[20]

Bernese Mittelland

The Bernese Mittelland (Bernese Midlands) is made up of the valley of the rivers Aare, the river Emme, some of the foothills of the Bernese Alps, as well as the plain around the capital Bern, and has many small farms and hilly forested regions with small to mid-sized towns scattered throughout. It is perhaps best known by foreigners and visitors for the Emmental. The classic Swiss cheese with holes Emmentaler comes from this region's forests and pastures, of hilly and low mountainous countryside in the 1000 to 2000 m (**unknown operator: u'['** to **unknown operator: u'['** ft) m range.

Three Lakes Region and Bernese Jura

In the north of the canton lies the predominantly French speaking Three Lakes Region (Seeland), concentrated around Lake Biel, Neuchâtel, and Murten, which rises from the plain up to the northernmost Swiss mountain chain of the Jura. This area has more relaxed geography, although still some lower mountains and some waterfalls, and large lakes.

Bernese Oberland

The Bernese Oberland (German: *Berner Oberland*) constitute the north side of the Bernese Alps and the west side of the Urner Alps within the canton of Bern. The highest mountain in the Bernese Alps is the Finsteraarhorn at 4274 metres (ft), but the best known mountains are Eiger, Mönch, and Jungfrau.

Staubbachfall

The well known hiking and ski resorts in the eastern Oberland are located around Interlaken and the Jungfrau, such as car-free Mürren and Wengen in the Lauterbrunnen valley, and Grindelwald. Further east, in the Haslital are the Aareschlucht and the town of Meiringen, famous for the fateful scene of Sherlock Holmes's 'death' at the hands of Professor Moriarty on the nearby Reichenbach Falls. In the Western Bernese Oberland there are many other resorts and small villages catering to visitors. These are accessed from the lake town of Thun, and the most notable of them are Kandersteg with the Oeschinensee and Adelboden. Further west is the Simmental with Lenk and Zweisimmen and the Saanenland with the famous resorts Gstaad and Saanen.

The whole area is very mountainous, with steep cliffs, many glaciers, and countless waterfalls. It is renowned for its scenic beauty and the charm of the small Swiss villages that dot the area. As a result of this, tourism is one of the main sources of income in the Bernese Oberland. The region also has an extensive train network as well as many cable cars and funiculars, with the highest train station in Europe at the Jungfraujoch and the longest gondola cableway in the world from Grindelwald to the Männlichen.

Mountains in the canton of Bern include:

Wetterhorn, painting by Joseph Anton Koch, 1824

- Finsteraarhorn – 4274 m (ft)
- Aletschhorn – 4190 m (ft)
- Jungfrau – 4158 m (ft)
- Mönch – 4107 m (ft)
- Schreckhorn – 4078 m (ft)
- Eiger – 3969 m (ft)
- Wetterhorn – 3698 m (ft)
- Blümlisalp – 3664 m (ft)
- Sustenhorn – 3503 m (ft)
- Wildhorn – 3248 m (ft)

Government

The Grand Council of Bern (German: *Grosser Rat* / French: *Grand conseil*) is the parliament of the canton of Bern. It consists of 160 representatives elected by proportional representation for four-year terms of office. The French-speaking part of the canton, the Bernese Jura, has 12 seats guaranteed and 3 seats are guaranteed for the French-speaking minority of the bilingual district of Biel/Bienne.

The Grand Council, the cantonal parliament

The Executive Council of Bern (German: *Regierungsrat* / French: *Conseil-éxecutif*) is the government of the canton of Bern. This seven-member collegial body is elected by the people for a period of four years. The cantonal constitution reserves one seat in the Executive Council for a French-speaking citizen from the Bernese Jura.

The canton has a two-tiered court system, consisting of district courts and a cantonal Supreme Court (German: *Obergericht*, French: *Cour suprême*). There is also an administrative court (German: *Verwaltungsgericht*; French: *Tribunal administratif*) as well as other specialised courts and judicial boards.[21]

Political subdivisions

On 1 January 2010, the 26 districts (*Amtsbezirke*) were combined into 10 new districts (*Verwaltungskreise*):[22]

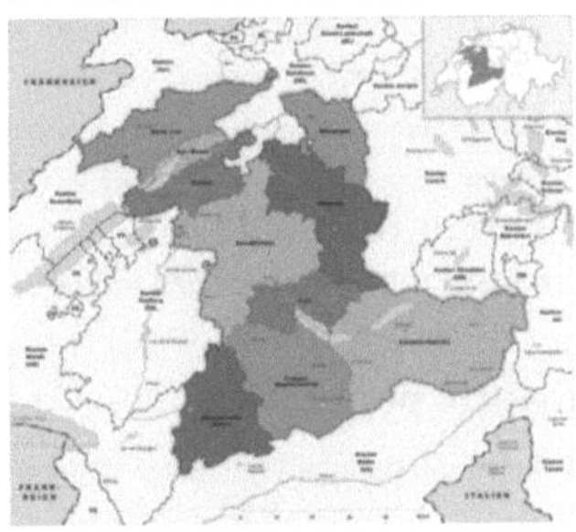
Districts of the canton of Bern

- Bern-Mittelland with capital Ostermundigen, made up of all or part of the former districts of Bern, Fraubrunnen, Konolfingen, Laupen, Schwarzenburg and Seftigen
- Biel/Bienne with capital Biel/Bienne, made up of all of the former district of Biel and about half of the former district of Nidau
- Emmental with capital Langnau im Emmental, made up of all or part of the former districts of Burgdorf, Signau and Trachselwald
- Frutigen-Niedersimmental with capital Frutigen, made up of all or part of the former districts of Frutigen and Niedersimmental
- Interlaken-Oberhasli with capital Interlaken, made up of all or part of the former districts of Interlaken and Oberhasli
- Jura bernois with capital Courtelary, made up of all or part of the former districts of Courtelary, Moutier and La Neuveville

- Oberaargau with capital Wangen an der Aare, made up of all or part of the former districts of Aarwangen and Wangen
- Obersimmental-Saanen with capital Saanen, made up of all of the former districts of Obersimmental and Saanen
- Seeland with capital Aarberg, made up of all or part of the former districts of Aarberg, Büren, Erlach and Nidau
- Thun with capital Thun, made up of all of the former district of Thun

Demographics

The canton of Bern is bilingual: Both German (22 districts, with 84% of the population) and French (three districts, 7.6%) are spoken.[20] The German-speaking majority speaks Bernese German, a Swiss German dialect. French-speakers live in the western and northern part of the canton. Both German and French are spoken in the bilingual city of Biel/Bienne. In the government and administration, both languages are official languages of equal standing.

Bern has a population (as of December 2011) of 985046.[1] As of 2010, 23.2% of the population are resident foreign nationals. Over the last 10 years (2000–2010) the population has changed at a rate of 3.3%. Migration accounted for 3.8%, while births and deaths accounted for –0.2%.[23] Most of the population (as of 2000) speaks German (804,190 or 84.0%) as their first language, French is the second most common (72,646 or 7.6%) and Italian is the third (18,908 or 2.0%). There are 688 people who speak Romansh.[24]

As of 2008, the population was 47.5% male and 52.5% female. The population was made up of 44,032 Swiss men (35.4% of the population) and 15,092 (12.1%) non-Swiss men. There were 51,531 Swiss women (41.4%) and 13,726 (11.0%) non-Swiss women.[25] Of the population in the canton, 292,559 or about 30.6% were born in Bern and lived there in 2000. There were 339,659 or 35.5% who were born in the same canton, while 154,709 or 16.2% were born somewhere else in Switzerland, and 129,864 or 13.6% were born outside of Switzerland.[24]

As of 2000, children and teenagers (0–19 years old) make up 22% of the population, while adults (20–64 years old) make up 60.9% and seniors (over 64 years old) make up 17.1%.[23]

As of 2000, there were 397,095 people who were single and never married in the canton. There were 449,014 married individuals, 61,206 widows or widowers and 49,882 individuals who are divorced.[24]

As of 2000, there were 415,901 private households in the canton, and an average of 2.2 persons per household.[23] There were 150,116 households that consist of only one person and 24,000 households with five or more people. As of 2009, the construction rate of new housing units was 4.3 new units per 1000 residents.[23]

As of 2003 the average price to rent an average apartment in Bern city was 1108.92 Swiss francs (CHF) per month (US$890, £500, €710 approx. exchange rate from 2003). The average rate for a one room apartment was 619.82 CHF (US$500, £280, €400), a two-room apartment was about 879.36 CHF (US$700, £400, €560), a three-room apartment was about 1040.54 CHF (US$830, £470, €670) and a six or more room apartment cost an average of 2094.80 CHF (US$1680, £940, €1340). The average apartment price in Bern was 99.4% of the national average of 1116 CHF.[26] The vacancy rate for the canton, in 2010, was 1.22%.[23]

Historic population

The historical population is given in the following chart:[27]

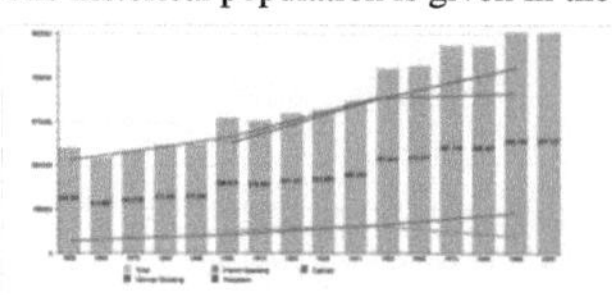

Politics

In the 2011 federal election the most popular party was the SVP which received 29.0% of the vote. The next three most popular parties were the SP/PS (19.3%), the BDP (14.9%) and the Green Party (9.4%).[28] The SVP received about the same percentage of the vote as they did in the 2007 Federal election (33.6% in 2007 vs 29.0% in 2011). The SPS retained about the same popularity (21.2% in 2007), the BDP moved from below fourth place in 2007 to third and the GPS moved from below fourth place in 2007 to fourth.[29]

Coat of arms

The blazon of the coat of arms is *Gules, on a bend or, a bear passant sable, langued, armed and vilené of the field.* The official blazon specifies that the tongue, claws and penis are red—and by extension it is important to always depict the bear as male.[30]

Religion

Most Bernese are Protestant (as of 2000, 67%),[] and most Protestants belong to the Swiss Reformed Church, which is officially recognised as a state church (German: *Landeskirche*), although it is autonomous in its governance and is organised along democratic principles. The canton is also home to a great number of small Evangelical Christian denominations unaffiliated with the state church. Bernese evangelical groups are mostly found in the Emmental and Berner Oberland, where they have a long tradition; several contemporary American religious groups, such as the Amish and Mennonites, were founded or co-founded by Bernese emigrants to the United States. Two small Evangelical political parties are represented in the Bernese cantonal parliament.

Capital city of Bern with river Aare, Gothic Nydeggchurch on left.

Bern features substantial Roman Catholic (16%)[] and Christian Catholic minorities. These churches also have state church status, and the small Jewish community is similarly recognised by law. As everywhere in Switzerland, there are also significant religious communities of immigrants, including Sikhs (who have a prominent Gurdwara, or temple, in Langenthal), The Church of Jesus Christ of Latter-day Saints (who have built the Bern Switzerland Temple) and Muslims. As of 2006, the plans to expand a backyard mosque in Langenthal with a symbolic minaret have, as elsewhere in Switzerland, caused a public stir due to vocal opposition from local conservative and evangelical leaders.[31]

From the 2000 census, 607,358 or 63.5% belonged to the Swiss Reformed Church, while 153,357 or 16.0% were Roman Catholic. Of the rest of the population, there were 9,153 members of an Orthodox church (or about 0.96% of the population), there were 1,064 individuals (or about 0.11% of the population) who belonged to the Christian Catholic Church, and there were 71,233 individuals (or about 7.44% of the population) who belonged to another Christian church. There were 807 individuals (or about 0.08% of the population) who were Jewish, and 28,377 (or about 2.96% of the population) who were Islamic. There were 2,662 individuals who were Buddhist, 5,991 individuals who were Hindu and 913 individuals who belonged to another church. 74,162 (or about 7.75% of the population) belonged to no church, are agnostic or atheist, and 37,059 individuals (or about 3.87% of the population) did not answer the question.[24]

Economy

Emmentaler cheese

Tourism is the main source of income in the Bernese Oberland. Other important sectors are agriculture (especially cattle breeding), cheese making, and hydroelectric power generation. The Bernese cheese Emmentaler is known around the world. In the Bernese Midlands the lands are more fertile. Agriculture is of great importance, but this part of the canton is also the most industrialized. Small and middle-sized businesses are important employers in this part of the canton of Bern. There is a nuclear power plant at Mühleberg.

The area around Lake Biel is renowned for its wine production. The 3 French-speaking districts of the Bernese Jura and the bilingual district of Biel/Bienne are renowned for their watch industry and its mechanical industry (high precision machine tools, automation and machining).

As of 2010, Bern had an unemployment rate of 2.8%. As of 2008, there were 36,685 people employed in the primary economic sector and about 12,638 businesses involved in this sector. 133,285 people were employed in the secondary sector and there were 10,141 businesses in this sector. 372,259 people were employed in the tertiary sector, with 34,818 businesses in this sector.[23] Of the working population, 25.6% used public transportation to get to work, and 43.4% used a private car.[23]

Education

In Bern about 385,640 or (40.3%) of the population have completed non-mandatory upper secondary education, and 121,749 or (12.7%) have completed additional higher education (either university or a *Fachhochschule*). Of the 121,749 who completed tertiary schooling, 65.0% were Swiss men, 24.8% were Swiss women, 6.2% were non-Swiss men and 4.0% were non-Swiss women.[24]

Notes and references

[1] Swiss Federal Statistical Office - STAT-TAB (http://www.pxweb.bfs.admin.ch/Database/German_01 - BevÃ¶lkerung/01.2 - BevÃ¶lkerungsstand und -bewegung/01.2 - BevÃ¶lkerungsstand und -bewegung.asp?lang=1&prod=01&secprod=2&openChild=true), online database – *Datenwürfel für Thema 01.2 - Bevölkerungsstand und -bewegung* (German) accessed 3 October 2012

[3] *Canton of Bern* in German (http://www.hls-dhs-dss.ch/textes/d/D7383.php), French (http://www.hls-dhs-dss.ch/textes/f/F7383.php) and Italian (http://www.hls-dhs-dss.ch/textes/i/I7383.php) in the online *Historical Dictionary of Switzerland.*

[4] UNESCO World Heritage Site – Prehistoric Pile dwellings around the Alps (http://whc.unesco.org/en/list/1363/)

[5] *Laupen* in German (http://www.hls-dhs-dss.ch/textes/d/D8447.php), French (http://www.hls-dhs-dss.ch/textes/f/F8447.php) and Italian (http://www.hls-dhs-dss.ch/textes/i/I8447.php) in the online *Historical Dictionary of Switzerland.*

[6] *von Kyburg* in German (http://www.hls-dhs-dss.ch/textes/d/D19520-1-1.php), French (http://www.hls-dhs-dss.ch/textes/f/F19520-1-1.php) and Italian (http://www.hls-dhs-dss.ch/textes/i/I19520-1-1.php) in the online *Historical Dictionary of Switzerland.*

[8] *Böser Bund in Berner Oberland* in German (http://www.hls-dhs-dss.ch/textes/d/D26842.php), French (http://www.hls-dhs-dss.ch/textes/f/F26842.php) and Italian (http://www.hls-dhs-dss.ch/textes/i/I26842.php) in the online *Historical Dictionary of Switzerland.*

[9] *Bernese Oberland* in German (http://www.hls-dhs-dss.ch/textes/d/D10296.php), French (http://www.hls-dhs-dss.ch/textes/f/F10296.php) and Italian (http://www.hls-dhs-dss.ch/textes/i/I10296.php) in the online *Historical Dictionary of Switzerland.*

[10] *Aarberg (Herrschaft, Amtsbezirk)* in German (http://www.hls-dhs-dss.ch/textes/d/D8434.php), French (http://www.hls-dhs-dss.ch/textes/f/F8434.php) and Italian (http://www.hls-dhs-dss.ch/textes/i/I8434.php) in the online *Historical Dictionary of Switzerland.*

[11] *Burgdorferkrieg* in German (http://www.hls-dhs-dss.ch/textes/d/D8733.php), French (http://www.hls-dhs-dss.ch/textes/f/F8733.php) and Italian (http://www.hls-dhs-dss.ch/textes/i/I8733.php) in the online *Historical Dictionary of Switzerland.*

[12] *Aigle* in German (http://www.hls-dhs-dss.ch/textes/d/D2276.php), French (http://www.hls-dhs-dss.ch/textes/f/F2276.php) and Italian (http://www.hls-dhs-dss.ch/textes/i/I2276.php) in the online *Historical Dictionary of Switzerland.*

[13] *Grandson district* in German (http://www.hls-dhs-dss.ch/textes/d/D7574.php), French (http://www.hls-dhs-dss.ch/textes/f/F7574.php) and Italian (http://www.hls-dhs-dss.ch/textes/i/I7574.php) in the online *Historical Dictionary of Switzerland.*

[14] *Interlaken Abbey and District* in German (http://www.hls-dhs-dss.ch/textes/d/D8506.php), French (http://www.hls-dhs-dss.ch/textes/f/F8506.php) and Italian (http://www.hls-dhs-dss.ch/textes/i/I8506.php) in the online *Historical Dictionary of Switzerland.*

[15] *Saanen district* in German (http://www.hls-dhs-dss.ch/textes/d/D8508.php), French (http://www.hls-dhs-dss.ch/textes/f/F8508.php) and Italian (http://www.hls-dhs-dss.ch/textes/i/I8508.php) in the online *Historical Dictionary of Switzerland*.
[16] *Château-d'Oex* in German (http://www.hls-dhs-dss.ch/textes/d/D2593.php), French (http://www.hls-dhs-dss.ch/textes/f/F2593.php) and Italian (http://www.hls-dhs-dss.ch/textes/i/I2593.php) in the online *Historical Dictionary of Switzerland*.
[17] *Histoire de la Suisse*, Éditions Fragnière, Fribourg, Switzerland.
[18] *Canton of Oberland* in German (http://www.hls-dhs-dss.ch/textes/d/D8633.php), French (http://www.hls-dhs-dss.ch/textes/f/F8633.php) and Italian (http://www.hls-dhs-dss.ch/textes/i/I8633.php) in the online *Historical Dictionary of Switzerland*.
[19] Kanton Bern, Regionen (http://www.vol.be.ch/site/wkb/wkb-regionen.htm) (german). Retrieved 30 July 2009
[20] Swiss Federal Statistical Office (http://www.bfs.admin.ch/bfs/portal/en/index/regionen/regionalportraets/bern/blank/kennzahlen.html) – Key Data. Retrieved 26 May 2009
[21] Gerichte und Strafverfolgungsbehörden (http://www.jgk.be.ch/site/index/g_gerichte.htm)
[22] Amtliches Gemeindeverzeichnis der Schweiz, Mutationsmeldungen 2009 / Répertoire officiel des communes de Suisse, Mutations 2009 / Elenco ufficiale dei Comuni della Svizzera, Mutazione 2009 (http://www.bfs.admin.ch/bfs/portal/de/index/infothek/nomenklaturen/blank/blank/gem_liste/03.Document.128095.pdf) (Report). Federal Statistical Office. 2009. nden. . Retrieved 6 March 2010.
[23] Swiss Federal Statistical Office (http://www.bfs.admin.ch/bfs/portal/en/index/regionen/regionalportraets.html). Retrieved 8 January 2012
[24] STAT-TAB Datenwürfel für Thema 40.3 – 2000 (http://www.pxweb.bfs.admin.ch/Database/German_40 - EidgenÃ¶ssische VolkszÃ¤hlung/40.3 - 2000/40.3 - 2000.asp?lang=1&prod=40&secprod=3&openChild=true) (German). Retrieved 2 February 2011
[25] Statistical office of the Canton of Bern (http://www.fin.be.ch/de/index/finanzen/finanzen/publikationen/wohnbevoelkerung.html) (German). Retrieved 4 January 2012
[26] Swiss Federal Statistical Office-Rental prices (http://www.bfs.admin.ch/bfs/portal/de/index/themen/05/06/blank/key/einfuehrung.html) 2003 data (German). Retrieved 26 May 2010
[27] Swiss Federal Statistical Office STAT-TAB Bevölkerungsentwicklung nach Region, 1850–2000 (http://www.pxweb.bfs.admin.ch/Database/German_40 - EidgenÃ¶ssische VolkszÃ¤hlung/40.4 - 1850-2000/40.4 - 1850-2000.asp?lang=1&prod=40&secprod=4&openChild=true) (German). Retrieved 29 January 2011
[28] Swiss Federal Statistical Office, Elections in Switzerland (http://www.portal-stat.admin.ch/nrw/files/de/01.xml) (German). Retrieved 5 January 2012
[29] Swiss Federal Statistical Office, *Nationalratswahlen 2007: Stärke der Parteien und Wahlbeteiligung, nach Gemeinden/Bezirk/Canton* (http://www.bfs.admin.ch/bfs/portal/de/index/themen/17/02/blank/data/04/03.html) (German). Retrieved 28 May 2010
[30] Flags of the World.com (http://www.crwflags.com/fotw/flags/ch-be.html). Retrieved 8 January 2012

See also

- Canton of Bern castles and fortresses
- List of mountains of the canton of Bern
- Berner Zeitung
- Der Bund
- Bieler Tagblatt

External links

- Berne Economic Development Agency (http://www.berneinvest.ch/3297/3512/3513.asp)
- Content in English and other languages (http://www.be.ch/web/index/english.htm) – Canton of Bern
- Canton of Bern (http://www.be.ch/) (German) (French)
- Official statistics (http://www.statistik.admin.ch/stat_ch/ber00/ekan_be.htm)
- Bernr Oberland Tourism (http://www.bernroberland.ch/navi/winter/frame_en.htm)
- Biel/Bienne and Seeland Tourism (http://www.biel-seeland.net/)
- Bernese Jura Tourism (http://www.jurabernois.ch/)
- *Canton of Bern* in German (http://www.hls-dhs-dss.ch/textes/d/D7383.php), French (http://www.hls-dhs-dss.ch/textes/f/F7383.php) and Italian (http://www.hls-dhs-dss.ch/textes/i/I7383.php) in the online *Historical Dictionary of Switzerland*.

Coat_of_arms

Escutcheon

Field

Supporter

Supporter

Motto (Scotland)

Crest

Torse

Mantling

Helm

Coronet

Compartment

Order

Ordinaries

Common

charges

Motto

Achievement elements

A **coat of arms** is a unique heraldic design on an escutcheon (i.e. shield), surcoat, or tabard. A surcoat, and subsequently a *coat of arms* was used by medieval knights to cover, protect, and identify the wearer. Thus these are sometimes termed *coat armory*. The coat of arms on an escutcheon forms the central element of the full heraldic achievement which consists of shield, supporters, crest, and motto. The design is a symbol unique to an individual person or family, corporation, or state. Such displays are commonly called *armorial bearings*, *armorial devices*, *heraldic devices*, or simply *arms*.

The ancient Romans used similar insignias similar on their shields, but these identified military units rather than individuals. The first evidence of medieval coats of arms is found in the Bayeux Tapestry from the 11th Century, where some of the combatants carry shields painted with crosses. Coats of arms came into general use by feudal lords and knights in battle in the 12th Century. By the 13th Century arms had spread beyond their initial battlefield use to become a kind of flag or logo for families in the higher social classes of Europe, inherited from one generation to the next. Exactly who had a right to use arms, by law or social convention, varied to some degree between countries. In the German-speaking region both the aristocracy and *burghers* (non-noble free citizens) used arms, while in most of the rest of Europe they were limited to the aristocracy. The use of arms spread to Church clergy, and to towns as civic identifiers, and to royally-chartered organizations such as universities and trading companies. Flags developed from coats of arms, and the arts of vexillology and heraldry are closely related. The coats of arms granted to commercial companies are a major source of the modern logo.

Despite no widespread regulation, and even with a lack in many cases of national regulation, heraldry has remained rather consistent across Europe, where traditions alone have governed the design and use of arms. Unlike seals and other general emblems, heraldic achievements have a formal description called a blazon, expressed in a jargon that allows for consistency in heraldic depictions.

In the 21st century, coats of arms are still in use by a variety of institutions and individuals; for example, many European cities and universities have guidelines on how their coats of arms may be used, and protect their use as trademarks.[1] [2] [3] Many societies exist that also aid in the design and registration of personal arms. Some nations, like England and Scotland, still maintain the same heraldic authorities which have traditionally granted and regulated arms for centuries and continue to do so in the present day.

Traditions and usage

In the heraldic traditions of England and Scotland an individual, rather than a family, had a coat of arms. In those traditions coats of arms are legal property transmitted from father to son; wives and daughters could also bear arms modified to indicate their relation to the current holder of the arms. Undifferenced arms are used only by one person at any given time. Other descendants of the original bearer could bear the ancestral arms only with some difference: usually a color change or the addition of a distinguishing charge. One such charge is the label, which in British usage (outside the Royal Family) is now always the mark of an heir apparent or (in Scotland) an heir presumptive.

The German *Hyghalmen Roll*, ca. late 15th century, illustrates the German practice of thematic repetition from the arms in the crest

Because of their importance in identification, particularly in seals on legal documents, the use of arms was strictly regulated; few countries continue in this today. This has been carried out by heralds and the study of coats of arms is therefore called "heraldry". Some other traditions (e.g., Polish heraldry) are less restrictive — allowing, for example, all members of a dynastic house or family to use the same arms, although one or more elements may be reserved to the head of the house.

In time, the use of arms spread from military entities to educational institutes, and other establishments.[2] According to a design institute article, "The modern logo and corporate livery have evolved from the battle standard and military uniform of medieval times".[4]

In his book, *The Visual Culture of Violence in the Late Middle Ages*, Valentin Groebner argues that the images composed on coats of arms are in many cases designed to convey a feeling of power and strength, often in military terms.[5] The author Helen Stuart argues that some coats of arms were a form of corporate logo.[6] Museums on medieval armory also point out that as emblems they may be viewed as precursors to the corporate logos of modern society, used for group identity formation.[7] [8]

When knights were so encased in armour that no means of identifying them was left, the practice was introduced of painting their insignia of honour on their shield as an easy method of distinguishing them. Originally these were granted only to individuals, but were afterward made hereditary in England by King Richard I, during his crusade to

Palestine.

European tradition

British heraldry

In Scotland, the Lord Lyon King of Arms has criminal jurisdiction to enforce the laws of arms. In England, Northern Ireland and Wales the use of arms is a matter of civil law and regulated by the College of Arms and the Court of Chivalry.

In reference to a dispute over the exercise of authority over the Officers of Arms in England, Arthur Annesley, 1st Earl of Anglesey, Lord Privy Seal, declared on 16 June 1673 that the powers of the Earl Marshal were "to order, judge, and determine all matters touching arms, ensigns of nobility, honour, and chivalry; to make laws, ordinances, and statutes for the good government of the Officers of Arms; to nominate Officers to fill vacancies in the College of Arms; to punish and correct Officers of Arms for misbehaviour in the execution of their places". It was further declared that no patents of arms or any ensigns of nobility should be granted and no augmentation, alteration, or addition should be made to arms without the consent of the Earl Marshal.

The coat of arms of the University of Plymouth, which were granted in 2008 by the College of Arms, shows the heater shield and style popular in modern British heraldry.[9]

Irish heraldry

In Ireland, since 1552, the usage and granting of coats of arms has been strictly regulated by the Government of Ireland, through the Office of the Chief Herald of Ireland.

German and Scandinavian heraldry

The heraldic tradition and style of modern and historic Germany and the Holy Roman Empire — including national and civic arms, noble and burgher arms, ecclesiastical heraldry, heraldic displays and heraldic descriptions — stand in contrast to Gallo-British, Latin and Eastern heraldry, and strongly influenced the styles and customs of heraldry in the Nordic countries, which developed comparatively late.[10]

Coat of arms of the city of Vaasa, showing the shield with the Royal House of Wasa emblem, a crown and a Cross of Liberty pendant.

In the Nordic countries, provinces, regions, cities and municipalities have a coat of arms. These are posted at the borders and on buildings containing official offices, as well as used in official documents and on the uniforms of municipal officers. Arms may also be used on souvenirs or other effects, given that an application has been granted by the municipal council.

Other European countries

At a national level, "coats of arms" were generally retained by European states with constitutional continuity of more than a few centuries, including constitutional monarchies like Denmark as well as old republics like San Marino and Switzerland.

In Italy the use of coats of arms was only loosely regulated by the states existing before the unification of 1860. Since the Consulta Araldica, the college of arms of the Kingdom of Italy, was abolished in 1948, personal coats of arms and titles of nobility, though not outlawed, are not recognised.

Among the states ruled by communist regimes, emblems resembling the Soviet design were adopted in all the Warsaw Pact states except Czechoslovakia and Poland.

Since 1989, some of the ex-Communist states, as Romania, have resumed their former arms, often with only the symbols of monarchy removed.

Asia and Africa

Arab world

Flag of the Ottoman Empire (1844–1924)

Egyptian coats of arms showing common Near and Middle Eastern motifs, namely the crescent and stars which are symbols of the region's predominant religion, Islam, and an eagle.

With the formation of the modern nation states of the Arab World in the second half of the 20th century, European traditions of heraldry were partially adopted for state emblems.

These emblems often involve the star and crescent symbol taken from the Ottoman flag. Another commonly seen charge is the eagle, which is a symbol attributed to Saladin, and the hawk of the Qureish.

Mitsubishi logo

Japanese Mon

Japanese emblems, called *kamon* (often abbreviated "mon"), are family badges which often date back to the 7th century, and are used in Japan today. The Japanese tradition is independent of the European, and thus very different in style; but as in Europe many abstract and floral elements are used.

Yet, even these simple designs often express an origin. An example in recent use is the logo of Mitsubishi corporation which started as a shipping and maritime enterprise and whose emblem is based on a water chestnut derived from its maritime history with a military naval influence. The word *mitsu* means the number 3 and the word *hishi* meaning "water chestnut" (pronounced *bishi* in some combinations; see rendaku) originated from the emblem of the warrior Tosa Clan. The battleships of the Tosa Clan had been used in the late 19th century in the First Sino-Japanese War to reach Korea[11] and their name was given to a modern battleship. The Tosa water chestnut leaf mon was then drawn as a rhombus or diamond shape in the Mitsubishi logo.[12] [13]

New World practices

Canada

The Queen of Canada has delegated her prerogative to grant armorial bearings to the Governor General of Canada. Canada has its own Chief Herald and Herald Chancellor. The Canadian Heraldic Authority is situated at Rideau Hall.[] []

United States

The Great Seal of the United States uses on the obverse as its central motif an heraldic achievement described as being the arms of the nation.[14] The seal, and the armorial bearings, were adopted by the Continental Congress on 20 June 1782, and is a shield divided palewise into thirteen pieces, with a blue chief, which is displayed upon the breast of an American bald eagle. The crest is thirteen stars breaking through a glory and clouds, displayed with no helm, torse or mantling (unlike most European precedents).

Only a few of the American states have adopted a coat of arms, which is usually designed as part of the respective state's seal. Vermont has both a state seal and a state coat of arms that are independent of one another (though both contain a pine tree, a cow and sheaves of grain); the seal is used to authenticate documents, whilst the heraldic device represents the state itself.

Ecclesiastic practice

Catholic Church

Vatican City State and the Holy See each have their own coat of arms. As the Papacy is not hereditary, its occupants display their personal arms combined with those of their office.

Some Popes came from armigerous (noble) families; others adopted coats of arms during their career in the Church. The latter typically allude to their ideal of life, or to specific Pontifical programmes.[15] A well-known and widely displayed example in recent times was Pope John Paul II's arms. His selection of a large letter M (for the Virgin Mary) was intended to express the message of his strong Marian devotion.[16]

Coat of arms of Pope John Paul II, display the Papal tiara and crossed keys of the pontifical office.

Roman Catholic dioceses also are assigned a coat of arms, as do a basilica or papal church also gets a coat of arms, the latter usually displaying these on the building. These may be used in countries which otherwise do not use heraldic devices.

In countries like Scotland with a strong statutory heraldic authority, arms will need to be officially granted and recorded.

Flags and banners

Flags are used to identify ships (where they are called ensigns), embassies and such, and they use the same colors and designs found in heraldry, but they are not usually considered to be heraldic. A country may have both a national flag and a national coat of arms, and the two may not look alike at all. For example, the flag of Scotland (St Andrew's Cross) has a white saltire on a blue field, but the royal arms of Scotland has a red lion within a double tressure on a gold (or) field.

See also

- National emblem
- Seal
- Insignia
- List of coats of arms
- Baron and feme
- Siebmachers Wappenbuch (Coats of arms from Germany, Switzerland, Austria, etc.)

Notes

[1] Glasgow University - The Coat of Arms (http://www.universitystory.gla.ac.uk/coat-of-arms/)
[2] Educational Institute Coat of arms (http://www.emeraldinsight.com/Insight/ViewContentServlet?Filename=Published/EmeraldFullTextArticle/Articles/1680100103.html)
[3] Policy on use of the Workmark and Insignia of McGill University (http://www.mcgill.ca/files/secretariat/WordmarkandInsigniaofMcGillUniversity-Policyontheuseof.pdf)
[4] Understanding Corporate Identity (http://www.dia.org.au/content.cfm?news=30&id=102)
[5] Valentin Groebner, 2004, *The Visual Culture of Violence in the Late Middle Ages* ISBN 978-1-890951-37-5
[6] *Employee Identification with the Corporate Identity* International Studies of Management and Organization, Volume 32, Number 3, 2002
[7] Group Identity Formation in the German Renaissance (http://www.phil-hum-ren.uni-muenchen.de/GermLat/Acta/Bernstein.htm)
[8] European Coat of Arms Museum (http://www.wien.gv.at/english/culture/museums/coat-of-arms.htm)
[9] College of Arms (http://www.college-of-arms.gov.uk/Newsletter/018.htm)
[10] ISBN 0-7137-0940-5 p.129.
[11] Donald Calman, 1992 *The Nature and Origins of Japanese Imperialism* ISBN 0-415-06710-3
[12] Mitsubishi Mark (http://www.mitsubishi.com/e/group/mark.html)
[13] Mitsubishi History (http://www.mitsubishi.com/e/history/index.html)
[14] 2004 Seal Broch (http://www.state.gov/documents/organization/27807.pdf)
[15] Coat of Arms of His Holiness Benedict XVI (http://www.vatican.va/holy_father/benedict_xvi/elezione/stemma-benedict-xvi_en.html)
[16] Vatican press office (http://www.vatican.va/news_services/press/documentazione/documents/sp_ss_scv/insigne/sp_ss_scv_stemma-bandiera-sigillo_en.html)

References

- This article incorporates text from a publication now in the public domain: "Pimbley's dictionary of heraldry", Arthur Francis Pimbley (1908)

External links

- College of Arms (http://www.college-of-arms.gov.uk/): Repository of the coats of arms and pedigrees of English, Welsh, Northern Irish and Commonwealth families and their descendants together with, and in principle under the control of, the legal body the Court of Chivalry, both medieval in origin.
- The Court of the Lord Lyon (http://www.lyon-court.com/lordlyon/CCC_FirstPage.jsp): the statutory heraldry office for Scotland
- The Armorial Register, International Register of Arms (http://www.armorial-register.com): A fully illustrated record of contemporary Coats of Arms
- Free access (http://www.luz-herald.net/free/) to Burke's General Armory (incomplete, 1500 British surnames), Pimbley's Dictionary of Heraldry and Blason des familles d'Europe, Grand Armorial Universel (15,000 European surnames)
- Royal Dutch Library page for the "Wapenboek Beyeren" (http://www.kb.nl/galerie/wapenboek/index.html/) written by Claes Heynenzoon around 1400, containing over 1000 drawings of coats of arms
- General armorial of noble families in the Russian Empire (Gerbovnik) (http://gerbovnik.com)

Agricultural_cooperative

An **agricultural cooperative**, also known as a **farmers' co-op**, is a cooperative where farmers pool their resources in certain areas of activity. A broad typology of agricultural cooperatives distinguishes between *agricultural service cooperatives*, which provide various services to their individually farming members, and *agricultural production cooperatives*, where production resources (land, machinery) are pooled and members farm jointly.[1] Examples of agricultural production cooperatives include collective farms in former socialist countries, the kibbutzim in Israel, collectively governed community shared agriculture, Longo Mai co-operatives [] and Nicaraguan production co-operatives.[2] Worker cooperatives provide an example of production cooperatives outside agriculture.

President of an agricultural cooperative in Guinea in an onion field

The default meaning of *agricultural cooperative* in English is usually an agricultural *service* cooperative, which is the numerically dominant form in the world. There are two primary types of agricultural service cooperatives, *supply cooperative* and *marketing cooperative*. Supply cooperatives supply their members with inputs for agricultural production, including seeds, fertilizers, fuel, and machinery services. Marketing cooperatives are established by farmers to undertake transportation, packaging, distribution, and marketing of farm products (both crop and livestock). Farmers also widely rely on credit cooperatives as a source of financing for both working capital and investments.

Why farmers form cooperatives

Cooperatives as a form of business organization are distinct from the more common investor-owned firms (IOFs).[1][3] Both are organized as corporations, but IOFs pursue profit maximization objectives, whereas cooperatives strive to maximize the benefits they generate for their members (which usually involves zero-profit operation). Agricultural cooperatives are therefore created in situations where farmers cannot obtain essential services from IOFs (because the provision of these services is judged to be unprofitable by the IOFs), or when IOFs provide the services at disadvantageous terms to the farmers (i.e., the services are available, but the profit-motivated prices are too high for the farmers). The former situations are characterized in economic theory as market failure or missing services motive. The latter drive the creation of cooperatives as a competitive yardstick or as a means of allowing farmers to build countervailing market power to oppose the IOFs.[1] The concept of competitive yardstick implies that farmers, faced with unsatisfactory performance by IOFs, may form a cooperative firm whose purpose is to force the IOFs, through competition, to improve their service to farmers.[3]

Headquarters of Hokuren Federation of Agricultural Cooperatives in Sapporo, Japan

A practical motivation for the creation of agricultural cooperatives is related to the ability of farmers to pool production and/or resources. In many situations within agriculture, it is simply too expensive for farmers to manufacture products or undertake a service. Cooperatives provide a method for farmers to join together in an 'association', through which a group of farmers can acquire a better outcome, typically financial, than by going alone. This approach is aligned to the concept of economies of scale and can also be related as a form of economic synergy, where "two or more agents working together to produce a result not obtainable by any of the agents independently". While it may seem reasonable to conclude that larger the cooperative the better, this is not necessarily true. Cooperatives exist across a broad membership base, with some cooperatives having less than 20 members while other can have over 10,000.

While the economic benefits are a strong driver in forming cooperatives, it is not the sole consideration. In fact, it is possible for the economic benefits from a cooperative to be replicated in other organisational forms, such as an IOF. An important strength of a cooperative for the farmer is that they retain the governance of the association, thereby ensuring they have ultimate ownership and control. This ensures that the profit reimbursement (either through the dividend payout or rebate) is shared only amongst the farmer members, rather than shareholders as in an IOF.

In agriculture, there are broadly three types of cooperatives: a machinery pool, a manufacturing/marketing cooperative, and a credit union.

- **Machinery Pool:** A family farm may be too small to justify the purchase of expensive farm machinery, which may be only used irregularly, say only during harvest; instead local farmers may get together to form a machinery pool that purchases the necessary equipment for all the members to use.
- **Manufacturing/marketing cooperative:** A farm does not always have the means of transportation necessary for delivering its produce to the market, or else the small volume of its production may put it in an unfavorable negotiating position with respect to intermediaries and wholesalers; a cooperative will act as an integrator, collecting the output from members, sometimes undertaking manufacturing, and delivering it in large aggregated quantities downstream through the marketing channels.
- **Credit Union:** Farmers, especially in developing countries, can be charged relatively high interest rates by commercial banks, or even not available for farmers to access. When providing loans, these banks are often mindful of high transaction costs on small loans, or may be refused credit altogether due to lack of collateral - something very acute in developing countries. To provide a source of credit, farmers can group together funds that can be loaned out to members. Alteratively, the credit union can raise loans at better rates from commercial banks due to the cooperative having a larger associative size than an individual farmer. Often members of a credit union will provide mutual or peer-pressure guarantees for repayment of loans. In some instances, manufacturing/marketing cooperatives may have credit unions as part of their broader business. Such an approach allows farmers to have a more direct access to critical farm inputs, such as seeds and implements. The loans for these inputs are repaid when the farmer sends produce to the manufacturing/marketing cooperative.

Supply cooperatives

Agricultural supply cooperatives aggregate purchases, storage, and distribution of farm inputs for their members. By taking advantage of volume discounts and utilizing other economies of scale, supply cooperatives bring down the cost of the inputs that the members purchase from the cooperative compared with direct purchases from commercial suppliers. Supply cooperatives provide inputs required for agricultural production including seeds, fertilizers, chemicals, fuel, and farm machinery. Some supply cooperatives operate machinery pools that provide mechanical field services (e.g., plowing, harvesting) to their members.

Examples

Canada

- Farmers' Storehouse Company
- United Farmers of Alberta
- Farmers of North America

Israel

- Granot central cooperative

Ukraine

- Ukrainian cooperative movement

United States

- Southern States Cooperative
- Farmers Cooperative Association, Inc.; Frederick, MD
- Ocean Spray (cooperative)

Netherlands

- Avebe
- Agrico
- Agrifirm

Marketing cooperatives

Agricultural marketing cooperatives are cooperative businesses owned by farmers, to undertake transformation, packaging, distribution, and marketing of farm products (both crop and livestock.)

Examples

New Zealand

New Zealand has a strong history in agricultural cooperatives, dating back to the late 19th century. The first was the small Otago Peninsula Co-operative Cheese Factory Co. Ltd, started in 1871 at Highcliff on the Otago Peninsula.[4][5] With active support by the New Zealand government, and small cooperatives being suitable in isolated areas, cooperatives quickly began to dominate the industry. By 1905, dairy cooperatives were the main organisational structure in the industry. In the 1920–30s, there were around 500 co-operative dairy companies compared to less than 70 that were privately owned.[6]

However, after World War II, with the advent of improved transportation, processing technologies and energy systems, a trend to merge dairy cooperatives occurred.[7] By the late 1990s, there were two major cooperatives: the Waikato-based New Zealand Dairy Group and the Taranaki-based Kiwi Co-operative Dairies. In 2001 these two cooperatives, together with the New Zealand Dairy Board, merged to form Fonterra. This mega-merger was supported by the New Zealand Government as part of broader dairy industry deregulation,[8] which allowed other companies to directly export dairy products. Two smaller cooperatives did not join Fonterra, preferring to remain independent – the Morrinsville-based Tatua Dairy Company and Westland Milk Products on the West Coast of the South Island.

The other main agricultural co-operatives in New Zealand are in the meat and fertiliser industries. The meat industry, which has struggled at times, has proposed various mergers similar to the creation of Fonterra, however these have failed to gain the necessary member support.[9]

Canada

In Canada, the most important cooperatives of this kind were the wheat pools. These farmer-owned cooperatives bought and transported grain throughout Western Canada. They replaced the earlier privately and often foreign-owned grain buyers and came to dominate the market in the post-war period. By the 1990s, most had demutualized (privatized), and several mergers occurred. Now all the former wheat pools are part of the Viterra corporation.

Former wheat pools include:

- Alberta Wheat Pool
- Manitoba Pool Elevators
- Saskatchewan Wheat Pool
- United Grain Growers

Other agricultural marketing cooperatives in Canada include:

- Organic Meadow Co-operative [10] (organic dairy)
- Gay Lea Foods Co-operative Limited (dairy)
- Agropur

India

In India, production of sugar from Sugarcane mostly takes place at Cooperative Sugar Mills owned by local Farmers The Shareholders include all farmers, small and large, supplying sugarcane to the mill.[11] Over the last sixty years, the local sugar mills have played a crucial part in encouraging rural political participation and as a stepping stone for aspiring politicians.[12] This is particularly true in the state of Maharashtra where a large number of politicians belonging to the Congress party or NCP had ties to sugar cooperatives from their respective local areas. Unfortunately, due to a policy of "profits for the company but losses to be borne by the government", has made a number of these operations inefficient.[13]

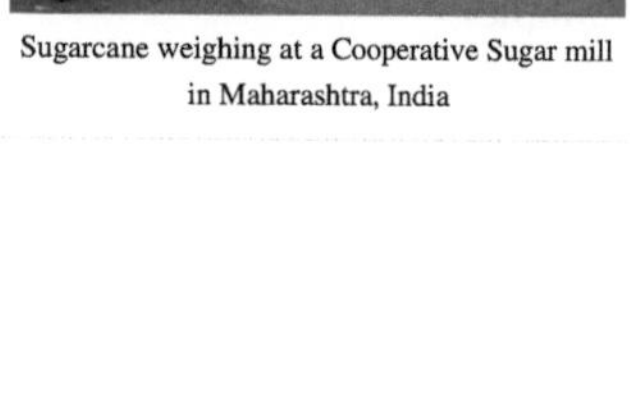

Sugarcane weighing at a Cooperative Sugar mill in Maharashtra, India

Israel

- Tnuva Central Cooperative for the Marketing of Agricultural Produce in Israel Ltd.

Netherlands

- CNB
- Coforta
- Royal Cosun
- ZON
- FloraHolland
- FrieslandCampina

Ukraine

- Ukrainian cooperative movement

United States

- American Legend Cooperative (mink fur) "Blackglama" brand
- Blue Diamond Growers (almonds)
- Cabot Creamery (dairy)
- Diamond of California (nuts), formerly a cooperative
- Dairylea Cooperative Inc. (Dairy), formerly Dairymen's League
- Dairy Farmers of America
- Florida's Natural Growers (citrus fruit)
- GreenStone Farm Credit Services [14] [financial products and services]
- Humboldt Creamery (dairy), formerly a cooperative
- Land O'Lakes (dairy and farm supply)
- Maine's Own Organic Milk Company (dairy)
- Michigan Sugar Company (sugar beets)
- Ocean Spray (cranberries and citrus fruit)
- Riceland Foods (rice, soybeans, and wheat)
- Snokist Growers (pears, apples, cherries)
- Sunkist Growers, Incorporated (citrus fruit)
- Sun-Maid (raisins)
- Sunsweet Growers Incorporated (dried fruit, especially prunes)
- Tillamook County Creamery Association (dairy)
- Lone Star Milk Producers (dairy)
- United Egg Producers

Mexico

- Zapatista coffee cooperatives

Production cooperatives

These are cooperative farms, jointly owned or managed by a cooperative society.

Cuba

Further information: CPA (agriculture)

Origins

The first agricultural cooperatives were created in Europe in the seventeenth century in the Military Frontier, where the wives and children of the border guards lived together in organized agricultural cooperatives next to a funfair and a public bath.

The first civil agricultural cooperatives were created also in Europe in the second half of the nineteenth century. They spread later to North America and the other continents. They have become one of the tools of agricultural development in emerging countries. Farmers also cooperated to form mutual farm insurance societies.

Also related are rural credit unions. They were created in the same periods, with the initial purpose of offering farm loans. Some became universal banks such as Crédit Agricole or Rabobank.

References

[1] Cobia, David, editor, *Cooperatives in Agriculture*, Prentice-Hall, Englewood Cliffs, NJ (1989), p. 50.

[2] Why Nicaraguan Peasants Stay in Agricultural Production Cooperatives (http://www.cedla.uva.nl/50_publications/pdf/revista/78RevistaEuropea/78_Ruben&Lerman.pdf) Ruerd Ruben and Zvi Lerman

[3] John M. Staatz, "Farmers' incentives to take collective action via cooperatives: A transaction-cost approach, " in: *Cooperative Theory: New Approaches*, ed. J.S. Royer, Washington, DC: USDA ACS Service Report 18 (July 1987), pp. 87-107.

[5] " Catherine Mathieson (http://www.dnzb.govt.nz/DNZB/alt_essayBody.asp?essayID=1M26)". *Dictionary of New Zealand Biography*.

[6] Philpott, H.G. (1937). *A History of the New Zealand Dairy Industry: 1840-1935*. Wellington: Government Printer.

[7] Ward, A.H. (1975). *A Command of Cooperatives*. Wellington: The New Zealand Dairy Board.

[8] Fonterra Co-Operative Group Ltd. (http://www.referenceforbusiness.com/history2/48/Fonterra-Co-Operative-Group-Ltd.html). Reference For Business.

[10] http://www.organicmeadow.com

[13] http://www.helsinki.fi/iehc2006/papers2/Das72.pdf

[14] http://www.greenstonefcs.com

Further Reading

- Derr, Jascha (2013), *The cooperative movement of Brazil and South Africa* (http://www.rosalux.co.za/wp-content/uploads/2013/04/Brazil_SA_Cooperatives-SD-1_2013.pdf)

See also

- Winemaking cooperative

Kerns

Kerns

Country Switzerland
Canton Obwalden
District n.a.

Population	5697 (Dec 2011)[1]
- Density	61 /km^2 (159 /sq mi)
Area	km^2 (sq mi)[2]
Elevation	564 m (ft)
Postal code	6064
SFOS number	1404
Surrounded by	Alpnach, Dallenwil (NW), Ennetmoos (NW), Hasliberg (BE), Innertkirchen (BE), Lungern, Sachseln, Sarnen, Wolfenschiessen (NW)
Website	www.kerns.ch [3] SFSO statistics [4]

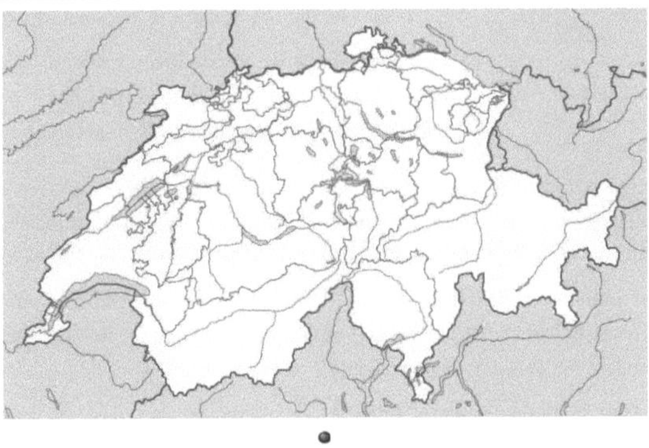

Kerns

Kerns is a village in the canton of Obwalden in Switzerland.

History

Kerns is first mentioned in 1036 as *Chernz* though this mention is from a 14th Century copy of the original document. After 1101 it was normally mentioned as *Chernes*.[4]

Geography

Swiss cow in the high meadows near Melchsee-Frutt

Kerns has an area, as of 2006, of 92.6 square kilometers (sq mi). Of this area, 46.7% is used for agricultural purposes, while 28.8% is forested. Of the rest of the land, 2.5% is settled (buildings or roads) and the remainder (21.9%) is non-productive (rivers, glaciers or mountains).[5]

The municipality is located on a terrace on the south-west slope of the Stanserhorn and Arvigrat mountains in the *Sarneraatal* and *Melchtal* valleys. It is the largest municipality in the canton. It consists of the village of Kerns with the sections of Dorf, Siebeneich, Wisserlen, Halten and Dietried, the hamlets of Sankt Niklausen and Melchtal as well as the resort of Melchsee-Frutt.

Demographics

Kerns has a population (as of 31 December 2011) of 5697.[1] As of 2007, 8.0% of the population was made up of foreign nationals.[6] Over the last 10 years the population has grown at a rate of 7.9%. Most of the population (as of 2000) speaks German (96.2%), with Albanian being second most common (1.3%) and Portuguese being third (0.5%).[5]

Chapel at Melchsee-Frutt with the Graustock mountain peak in the background

As of 2000 the gender distribution of the population was 49.8% male and 50.2% female. As of 2000 there are 1,700 households,[6] and in 2008 there were 2,110 households in Kerns alone, with 15 in St. Niklausen, 180 in Melchtal and 15 in Melchsee-Frutt.[7]

In the 2007 federal election the most popular party was the SVP which received 38.1% of the vote. The next three most popular parties were the CVP (36.6%), the Other (16.9%) and the SPS (8.3%).[5]

In Kerns about 67.5% of the population (between age 25-64) have completed either non-mandatory upper secondary education or additional higher education (either university or a *Fachhochschule*).[5]

Kerns has an unemployment rate of 0.85%. As of 2005, there were 445 people employed in the primary economic sector and about 172 businesses involved in this sector. 644 people are employed in the secondary sector and there are 65 businesses in this sector. 853 people are employed in the tertiary sector, with 136 businesses in this sector.[5]

As of 2007, the religious population distribution was; 4,654 or 83.6% are Roman Catholic, while 251 or 4.5% belonged to the Swiss Reformed Church. There are 660 individuals (or about 11.86% of the population) who belong to another church (not listed on the census), or no church.[7]

The historical population is given in the following table:[4] [7]

year	population
1744	1,820
1758	1,695
1799	1,999
1811	2,236
1850	2,509
1900	2,392
1950	3,406
1960	3,553
1970	3,881
1980	4,200
2000	5,101
2007	5,523

Sights

The main sights of Kerns are: chapel *St. Niklausen*, an old bridge (*Hohe Brücke*), and the pilgrimage church *Melchtal*. The ski area and mountain resort Melchsee-Frutt with the lake Melchsee is also part of the municipality of Kerns. The reservoir Tannensee is located above the village, at 1976 m (ft).

Hohe Brücke (High Bridge) near Kerns, picture from 1893

References

[1] Swiss Federal Statistics Office – STAT-TAB (http://www.pxweb.bfs.admin.ch/Dialog/varval.asp?ma=px-d-01-2A01&ti=Ständige+und+Nichtständige+Wohnbevölkerung+nach+Region,+Geschlecht,+Nationalität+und++Alter&path=../Database/German_01 - Bevölkerung/01.2 - Bevölkerungsstand und -bewegung/&lang=1&prod=01&openChild=true&secprod=2) *Ständige und Nichtständige Wohnbevölkerung nach Region, Geschlecht, Nationalität und Alter* (German) accessed 3 October 2012

[2] Arealstatistik Standard - Gemeindedaten nach 4 Hauptbereichen (http://www.bfs.admin.ch/bfs/portal/de/index/themen/02/03/blank/data/gemeindedaten.html)

[3] http://www.kerns.ch

[4] *Kerns* in German (http://www.hls-dhs-dss.ch/textes/d/D743.php), French (http://www.hls-dhs-dss.ch/textes/f/F743.php) and Italian (http://www.hls-dhs-dss.ch/textes/i/I743.php) in the online *Historical Dictionary of Switzerland*.

[5] Swiss Federal Statistical Office (http://www.bfs.admin.ch/bfs/portal/en/index/regionen/02/key.html) accessed 03-Sep-2009

[6] Canton Obwalden Statistics (http://www.statistik-obwalden.ch/index.cfm?navid=78&docid=78) accessed 2 September 2009 (German)

[7] Kerns Municipality website (http://www.kerns.ch/de/portrait/gemeindeinzahlen/)-Kerns in numbers. (German) accessed 3 September 2009

External links

- *Kerns* in German (http://www.hls-dhs-dss.ch/textes/d/D743.php), French (http://www.hls-dhs-dss.ch/textes/f/F743.php) and Italian (http://www.hls-dhs-dss.ch/textes/i/I743.php) in the online *Historical Dictionary of Switzerland*.

Schattenhalb

Schattenhalb	
Country	Switzerland
Canton	Bern
District	Interlaken-Oberhasli

Population	577 (Dec 2011)[1]
- Density	18 /km^2 (47 /sq mi)
Area	km^2 (sq mi)[2]
Elevation	600 m (ft)
Postal code	3860
SFOS number	0786
Mayor	Niklaus Leuenberger
Surrounded by	Meiringen, Innertkirchen and Grindelwald
Website	www.schattenhalb.ch [3] SFSO statistics [4]

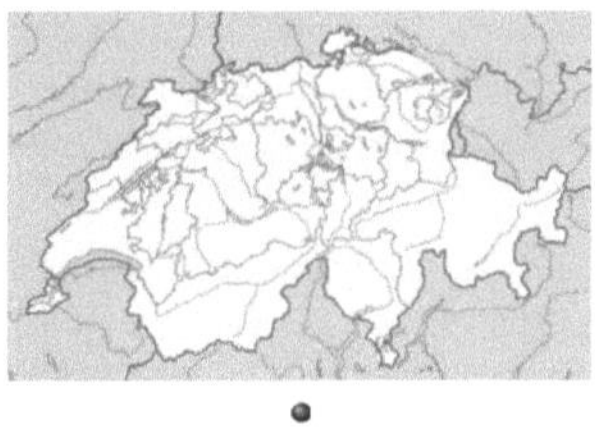

Schattenhalb

Schattenhalb is a municipality in the Interlaken-Oberhasli administrative district in the canton of Bern in Switzerland. The municipality includes the settlements of **Willigen**, **Geissholz**, **Falcheren** and **Lüögen**.[4]

History

The municipality was created in 1834 from the cooperative farms (*Bäuerten*) of "Willigen," "Geissholz," "Falchern" and "Lugen." The name Schattenhalb ("Shadow-half") refers to the shady face of the mountainside south of Meiringen.

Geography

Spa-hotel Rosenlaui

Schattenhalb has an area of 31.5 km^2 (sq mi). Of this area, 26.8% is used for agricultural purposes, while 39% is forested. Of the rest of the land, 1.7% is settled (buildings or roads) and the remainder (32.5%) is non-productive (rivers, glaciers or mountains).[5]

Schattenhalb lies in the Bernese Oberland in the Hasli valley, where the Aare River flows. The highest mountain is the Wellhorn, on the southwestern edge of the municipality (3,192 m).

Part of the municipality is covered by the Rosenlaui Glacier. At a lower altitude is the Rosenlaui gorge, a tourist attraction, which can be reached from the Hasli valley via the Reichenbachtal valley. At the lower end of this valley are the famous Reichenbach Falls.

Demographics

Schattenhalb has a population (as of 31 December 2011) of 577.[1] As of 2007, 11.3% of the population was made up of foreign nationals. Over the last 10 years the population has decreased at a rate of -1.3%. Most of the population (as of 2000) speaks German (92.8%), with Spanish being second most common (2.9%) and Portuguese being third (1.4%).

In the 2007 election the most popular party was the SVP which received 56.3% of the vote. The next three most popular parties were the SPS (17.5%), the FDP (12.5%) and the Green Party (7.7%).

The age distribution of the population (as of 2000) is children and teenagers (0–19 years old) make up 22.5% of the population, while adults (20–64 years old) make up 55.8% and seniors (over 64 years old) make up 21.7%. In Schattenhalb about 68.3% of the population (between age 25-64) have completed either non-mandatory upper secondary education or additional higher education (either university or a *Fachhochschule*).

Schattenhalb has an unemployment rate of 0.72%. As of 2005, there were 47 people employed in the primary economic sector and about 22 businesses involved in this sector. 92 people are employed in the secondary sector and there are 7 businesses in this sector. 448 people are employed in the tertiary sector, with 16 businesses in this sector.[5]

References

[1] Swiss Federal Statistical Office - STAT-TAB (http://www.pxweb.bfs.admin.ch/Database/German_01 - BevÃ¶lkerung/01.2 - BevÃ¶lkerungsstand und -bewegung/01.2 - BevÃ¶lkerungsstand und -bewegung.asp?lang=1&prod=01&secprod=2&openChild=true), online database – *Datenwürfel für Thema 01.2 - Bevölkerungsstand und -bewegung* (German) accessed 3 October 2012

[2] Arealstatistik Standard - Gemeindedaten nach 4 Hauptbereichen (http://www.bfs.admin.ch/bfs/portal/de/index/themen/02/03/blank/data/gemeindedaten.html)

[3] http://www.schattenhalb.ch

[4] Swiss Confederation. *map.geo.admin.ch* (http://map.geo.admin.ch/?selectedNode=node_ch.swisstopo.swissboundaries3d-kanton-flaeche.fill1&Y=656964&X=173874&zoom=7&bgLayer=ch.swisstopo.pixelkarte-farbe&layers=ch.swisstopo.swissboundaries3d-gemeinde-flaeche.fill&layers_opacity=1&layers_visibility=true&lang=en) (Map). . Retrieved 2013-01-05.

[5] Swiss Federal Statistical Office (http://www.bfs.admin.ch/bfs/portal/en/index/regionen/02/key.html) accessed 10-Jul-2009

Wolfenschiessen

Wolfenschiessen

Trübsee

Country	Switzerland
Canton	Nidwalden
District	n.a.

Population	2046 (Dec 2011)[1]
- Density	22 /km^2 (57 /sq mi)
Area	km^2 (sq mi)[2]
Elevation	510 m (ft) Wolfenschiessen village
• Highest	2880 m - Titlis Glacier
• Lowest	488 m - Niederrickenbach Station
Postal code	6386
SFOS number	1511
Surrounded by	Beckenried, Dallenwil, Engelberg (OW), Innertkirchen (BE), Isenthal (UR), Kerns (OW), Oberdorf
Website	www.wolfenschiessen.ch [3] SFSO statistics [4]

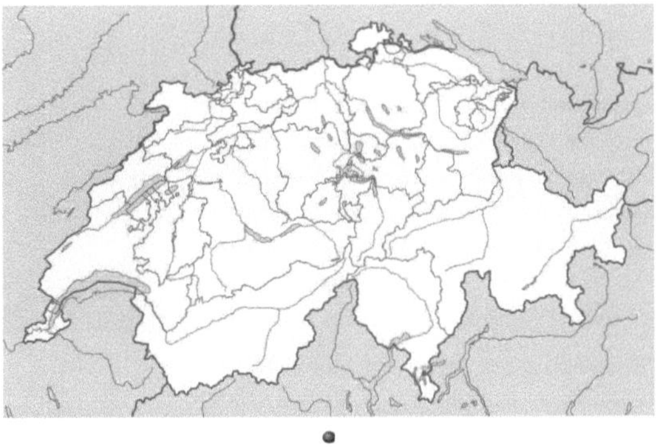

Wolfenschiessen

Wolfenschiessen is a village and municipality in the canton of Nidwalden in Switzerland. Besides the village of Wolfenschiessen itself, the municipality includes the settlements of **Altzellen**, **Büren ob dem Bach**, **Dörfli** and **Oberrickenbach**, together with a large area of high alpine land, mountains, lakes and glaciers.[4]

Coat of arms

The municipality's coat of arms is *Azure, a Wolf rampant Argent pierced by an Arrow in bend sinister Or*. This is an example of canting where the name of the municipality is translated or represented on the coat of arms. In this case, Wolfenschiessen means "to shoot the wolf."[5]

Geography

Wolfenshiessen encompasses both sides of the deep valley of the Engelberger Aa river downstream of Engelberg, with high alpine areas encircling Engelberg to the north, west and south. The settlements of Wolfenschiessen, Büren ob dem Bach and Dörfli lie in the valley, whilst Altzellen and Oberrickenbach are on higher ground to the east.[4]

Looking down from the summit of Mount Titlis over the Trübsee into the valley of the Engelberger Aa. Wolfenschiessen village is in the deepest part of that valley, to the top left corner of the image.

The lake of Bannalpsee lies above Oberrickenbach to the north of Engelberg, whilst Trübsee is on high ground to the south of Engelberg. In its far southern corner, the municipality reaches high up the flank of Mount Titlis, including part of the Titlis Glacier (*Titlisgletscher*).[4]

Wolfenschiessen has an area, as of 2006, of 92.8 km^2 (sq mi). Of this area, 40.8% is used for agricultural purposes, while 33% is forested. Of the rest of the land, 1.2% is settled (buildings or roads) and the remainder (24.9%) is non-productive (rivers, glaciers or mountains).[6]

Demographics

The Bannalpsee

Wolfenschiessen has a population (as of 31 December 2011) of 2046.[1] As of 2007, 9.7% of the population was made up of foreign nationals.[7] Over the last 10 years the population has grown at a rate of 0.3%. Most of the population (as of 2000) speaks German (93.6%), with Albanian being second most common (2.7%) and Serbo-Croatian being third (1.4%).[6] As of 2008 the gender distribution of the population was 51.7% male and 48.3% female.

As of 2000 there are 674 households, of which 348 households (or about 51.6%) contain only one or two individuals. 111 or about 16.5% are large households, with at least five members.[8]

In the 2007 federal election the most popular party was the FDP which received 89.3% of the vote. The next two most popular parties were the local, small right-wing parties (10.7%) and the SPS (0.1%).[6]

In Wolfenschiessen about 58.6% of the population (between age 25-64) have completed either non-mandatory upper secondary education or additional higher education (either university or a *Fachhochschule*).[6]

Wolfenschiessen has an unemployment rate of 0.88%. As of 2005, there were 228 people employed in the primary economic sector and about 94 businesses involved in this sector. 138 people are employed in the secondary sector and there are 26 businesses in this sector. 268 people are employed in the tertiary sector, with 61 businesses in this sector.[6]

The historical population is given in the following table:[8]

year	population
1960	1,647
1970	1,470
1980	1,524
1990	1,765
2000	2,041
2005	1,999

Transport

The municipality of Dallenwil is served by three railway stations on the Luzern–Stans–Engelberg line. Besides Wolfenschiessen station itself, there is Niederrickenbach station to the north and Dörfli station to the south. The hourly InterRegio service between Lucerne and Engelberg stops at all three stations.[9]

A post bus service within the municipality links the villages of Wolfenschiessen and Oberrickenbach, with eight services per day. From Oberrickenbach, an cable car rises to the Bannalpsee.[4] [10]

The lake of Trübsee is an intermediate stop on the cable car system that links Engelberg and the summit of Mount Titlis.[4]

Cable car from Oberrickenbach to the Bannalpsee

Sights

The main sights of Wolfenschiessen are: the parish church *St. Maria*, the chapel *St. Joder* in *Altzellen*, the mansions *Hechhuis*, and the *Grossitz* mansion.

References

[1] Swiss Federal Statistics Office – STAT-TAB (http://www.pxweb.bfs.admin.ch/Dialog/varval.asp?ma=px-d-01-2A01&ti=Ständige+und+Nichtständige+Wohnbevölkerung+nach+Region,+Geschlecht,+Nationalität+und++Alter&path=../Database/German_01 - Bevölkerung/01.2 - Bevölkerungsstand und -bewegung/&lang=1&prod=01&openChild=true&secprod=2) *Ständige und Nichtständige Wohnbevölkerung nach Region, Geschlecht, Nationalität und Alter* (German) accessed 3 October 2012

[2] Arealstatistik Standard - Gemeindedaten nach 4 Hauptbereichen (http://www.bfs.admin.ch/bfs/portal/de/index/themen/02/03/blank/data/gemeindedaten.html)

[3] http://www.wolfenschiessen.ch

[4] Swiss Confederation. *map.geo.admin.ch* (http://map.geo.admin.ch/?selectedNode=node_ch.swisstopo.swissboundaries3d-kanton-flaeche.fill1&Y=672885.5&X=195627&zoom=8&bgLayer=ch.swisstopo.pixelkarte-farbe&layers=ch.swisstopo.swissboundaries3d-gemeinde-flaeche.fill&layers_opacity=1&layers_visibility=true&lang=en) (Map). . Retrieved 2013-01-08.

[5] Flags of the World-Wolfenschiessen (http://flagspot.net/flags/ch-nw011.html) accessed 4 September 2009

[6] Swiss Federal Statistical Office (http://www.bfs.admin.ch/bfs/portal/en/index/regionen/02/key.html) accessed 04-Sep-2009

[7] Nidwalden Statistical Office-Population (http://www.nw.ch/de/verwaltung/departemente/welcome.php?departement_id=12&page=5&action=download&id=2575) (German) accessed 4 September 2009

[8] Nidwalden Statistical Office-Municipalities (http://www.nw.ch/de/onlinemain/publikationen/welcome.php?action=download&amt_id=&id=2200) (German) accessed 4 September 2009

Guttannen

Guttannen	
Country Switzerland **Canton** Bern **District** Interlaken-Oberhasli	
Population	300 (Dec 2011)[1]
- Density	1 /km^2 (4 /sq mi)
Area	km^2 (sq mi)[2]
Elevation	1057 m (ft)
Postal code	3864
SFOS number	0782
Mayor	Esther Messerli
Surrounded by	Fieschertal (VS), Gadmen, Grindelwald, Innertkirchen, Münster-Geschinen (VS), Obergesteln (VS), Oberwald (VS), Ulrichen (VS)
Website	www.guttannen.ch [3] SFSO statistics [4]

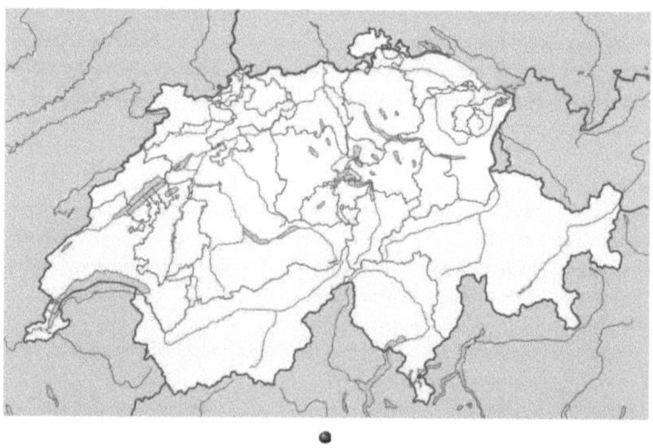

Guttannen

Guttannen is a municipality in the Interlaken-Oberhasli administrative district in the canton of Bern in Switzerland.

History

Guttannen is first mentioned in 1377 as *Guotentannon*.[4]

Geography

View of Lake Grimsel, Grimsel Hospice, and Lake Räterichsboden from a Grimsel Pass

Guttannen lies in the Bernese Oberland near Grimsel Pass. It is the highest settlement in the Haslital. The adjacent municipalities from the north clockwise are Innertkirchen, Gadmen, Oberwald, Obergesteln, Ulrichen, Münster-Geschinen, Fieschertal and Grindelwald.

The Aare springs from glaciers in Guttannen. There are four lakes in the municipality: Lake Oberaar, Lake Grimsel, Lake Räterichsboden and Lake Gelmer. In the western part of the municipality are the mountains Schreckhorn, Lauteraarhorn and Finsteraarhorn. Guttannen also includes Finsteraar Glacier, Lauteraar Glacier, Unteraar Glacier, Grueben Glacier, and Bächli Glacier.

Guttannen has an area, as of 2009, of 200.79 km^2 (sq mi). Of this area, 7.67 km^2 (sq mi) or 3.8% is used for agricultural purposes, while 18.77 km^2 (sq mi) or 9.3% is forested. Of the rest of the land, 0.98 km^2 (sq mi) or 0.5% is settled (buildings or roads), 7.45 km^2 (sq mi) or 3.7% is either rivers or lakes and 165.86 km^2 (sq mi) or 82.6% is unproductive land.[5]

Of the built up area, housing and buildings made up 0.1% and transportation infrastructure made up 0.3%. 3.8% of the total land area is heavily forested. Of the agricultural land, 2.9% is used for alpine pastures. All the water in the municipality is in lakes. Of the unproductive areas, 13.7% is unproductive vegetation, 45.6% is too rocky for vegetation and 23.3% of the land is covered by glaciers.[5]

By Swiss standards, its area is colossal; it is larger than the entire Canton of Appenzell Innerrhoden. It is also an evangelical-reformed parish.

Demographics

Lake Grimsel

Guttannen has a population (as of 31 December 2011) of 300.[1] As of 2007, 4.7% of the population was made up of foreign nationals. Over the last 10 years the population has decreased at a rate of -13.2%. Most of the population (as of 2000) speaks German (98.2%), with French being second most common (0.3%) and Rhaeto-romance being third (0.3%).

In the 2007 election the most popular party was the SVP which received 54.8% of the vote. The next three most popular parties were the FDP (15.7%), the SPS (12%) and the Green Party (10.4%).

The age distribution of the population (as of 2000) is children and teenagers (0–19 years old) make up 23.8% of the population, while adults (20–64 years old) make up 58.8% and seniors (over 64 years old) make up 17.4%. The entire Swiss population is generally well educated. In Guttannen about 64% of the population (between age 25-64) have completed either non-mandatory upper secondary education or additional higher education (either university or a *Fachhochschule*).

Guttannen has an unemployment rate of 1.05%. As of 2005, there were 83 people employed in the primary economic sector and about 27 businesses involved in this sector. 67 people are employed in the secondary sector and there are 8 businesses in this sector. 98 people are employed in the tertiary sector, with 12 businesses in this sector.[6] The historical population is given in the following table:[4]

year	population
1764	288
1850	506
1900	345
1920	298
1950	557
2000	328

Origin of the name

Guttannen is the name of a meadow, which became the name of the municipality. It comes from the phrase *ze den guoten tannen* (by the good firs).

References

[1] Swiss Federal Statistical Office - STAT-TAB (http://www.pxweb.bfs.admin.ch/Database/German_01 - BevÃ¶lkerung/01.2 - BevÃ¶lkerungsstand und -bewegung/01.2 - BevÃ¶lkerungsstand und -bewegung.asp?lang=1&prod=01&secprod=2&openChild=true), online database – *Datenwürfel für Thema 01.2 - Bevölkerungsstand und -bewegung* (German) accessed 3 October 2012

[2] Arealstatistik Standard - Gemeindedaten nach 4 Hauptbereichen (http://www.bfs.admin.ch/bfs/portal/de/index/themen/02/03/blank/data/gemeindedaten.html)

[3] http://www.guttannen.ch

[4] *Guttannen* in German (http://www.hls-dhs-dss.ch/textes/d/D468.php), French (http://www.hls-dhs-dss.ch/textes/f/F468.php) and Italian (http://www.hls-dhs-dss.ch/textes/i/I468.php) in the online *Historical Dictionary of Switzerland.*

[5] Swiss Federal Statistical Office-Land Use Statistics (http://www.bfs.admin.ch/bfs/portal/de/index/themen/02/03/blank/data/gemeindedaten.html) 2009 data (German) accessed 25 March 2010

[6] Swiss Federal Statistical Office (http://www.bfs.admin.ch/bfs/portal/en/index/regionen/02/key.html) accessed 10-Jul-2009

External links

- *Guttannen* in German (http://www.hls-dhs-dss.ch/textes/d/D468.php), French (http://www.hls-dhs-dss.ch/textes/f/F468.php) and Italian (http://www.hls-dhs-dss.ch/textes/i/I468.php) in the online *Historical Dictionary of Switzerland.*

Gadmen

Gadmen	
 Valley of Gadmen	
Country Switzerland **Canton** Bern **District** Interlaken-Oberhasli	
Population	228 (Dec 2011)[1]
- Density	2 /km^2 (5 /sq mi)
Area	km^2 (sq mi)[2]
Elevation	1205 m (ft)
Postal code	3863
SFOS number	0781
Mayor	Barbara Kehrli
Surrounded by	Engelberg (OW), Göschenen (UR), Guttannen, Innertkirchen, Oberwald (VS), Wassen (UR)
Twin towns	Volketswil (Switzerland)
Website	www.gadmen.ch [3] SFSO statistics [4]

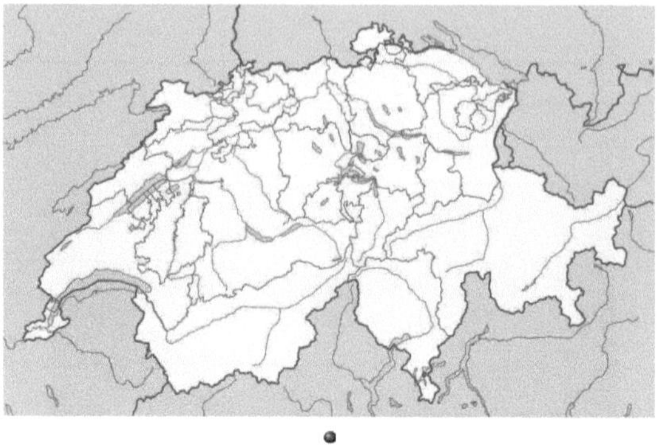

Gadmen

Gadmen is a municipality in the Interlaken-Oberhasli administrative district in the canton of Bern in Switzerland.

The name Gadmen comes from the Old High German "gadum," meaning barn or small house, and was first mentioned in 1382.

The municipality is in a municipal partnership with Volketswil in the Canton of Zürich.

History

Gadmen is first mentioned in 1382 as *im Gadmen*.[4]

Demographics

View of Steinsee and Gadmen from the Susten Pass

Gadmen has a population (as of 31 December 2011) of 228.[1] As of 2007, 2.8% of the population was made up of foreign nationals. Over the last 10 years the population has decreased at a rate of -15.5%. Most of the population (as of 2000) speaks German (98.5%), with French being second most common (0.4%) and Spanish being third (0.4%).

In the 2007 election the most popular party was the SVP which received 52.4% of the vote. The next three most popular parties were the SPS (22.1%), the FDP (13.6%) and the local small left-wing parties (4.6%).

The age distribution of the population (as of 2000) is children and teenagers (0–19 years old) make up 21.3% of the population, while adults (20–64 years old) make up 53.7% and seniors (over 64 years old) make up 25%. The entire Swiss population is generally well educated. In Gadmen about 51.1% of the population (between age 25-64) have completed either non-mandatory upper secondary education or additional higher education (either university or a *Fachhochschule*).

Gadmen has an unemployment rate of 1.12%. As of 2005, there were 52 people employed in the primary economic sector and about 21 businesses involved in this sector. 13 people are employed in the secondary sector and there are 3 businesses in this sector. 43 people are employed in the tertiary sector, with 13 businesses in this sector.[5] The historical population is given in the following table:[4]

year	population
1764	354
1850	739
1900	672
1950	485
1960	510
2000	272

Geography

Gadmen has an area, as of 2009, of 116.49 km^2 (sq mi). Of this area, 9.62 km^2 (sq mi) or 8.3% is used for agricultural purposes, while 19.48 km^2 (sq mi) or 16.7% is forested. Of the rest of the land, 0.75 km^2 (sq mi) or 0.6% is settled (buildings or roads), 1.52 km^2 (sq mi) or 1.3% is either rivers or lakes and 85.19 km^2 (sq mi) or 73.1% is unproductive land.[6]

Of the built up area, housing and buildings made up 0.1% and transportation infrastructure made up 0.4%. 10.9% of the total land area is heavily forested. Of the agricultural land, 1.7% is pastures and 6.5% is used for alpine pastures. All the water in the municipality is in rivers and streams. Of the unproductive areas, 13.3% is unproductive vegetation, 33.0% is too rocky for vegetation and 26.8% of the land is covered by glaciers.[6]

Gadmen lies in the Bernese Oberland, and the Susten Pass lies on the eastern border of the municipality. It lies in the Gadmer watershed. The neighboring municipalities are Engelberg, Wassen, Göschenen, Guttannen, Innertkirchen and Kerns.

There are three glaciers in Gadmen, the Stein Glacier, the Trift Glacier and the Wenden Glacier. Beneath the Stein Glacier is Lake Stein. The most famous mountain in Gadmen is the Titlis.

Economy

The largest industries are tourism and agriculture.

External links

- *Gadmen* in German [7], French [8] and Italian [9] in the online *Historical Dictionary of Switzerland.*

References

[1] Swiss Federal Statistical Office - STAT-TAB (http://www.pxweb.bfs.admin.ch/Database/German_01 - BevÃ¶lkerung/01.2 - BevÃ¶lkerungsstand und -bewegung/01.2 - BevÃ¶lkerungsstand und -bewegung.asp?lang=1&prod=01&secprod=2&openChild=true), online database – *Datenwürfel für Thema 01.2 - Bevölkerungsstand und -bewegung* (German) accessed 3 October 2012

[2] Arealstatistik Standard - Gemeindedaten nach 4 Hauptbereichen (http://www.bfs.admin.ch/bfs/portal/de/index/themen/02/03/blank/data/gemeindedaten.html)

[3] http://www.gadmen.ch

[4] *Gadmen* in German (http://www.hls-dhs-dss.ch/textes/d/D467.php), French (http://www.hls-dhs-dss.ch/textes/f/F467.php) and Italian (http://www.hls-dhs-dss.ch/textes/i/I467.php) in the online *Historical Dictionary of Switzerland.*

[5] Swiss Federal Statistical Office (http://www.bfs.admin.ch/bfs/portal/en/index/regionen/02/key.html) accessed 10-Jul-2009

[6] Swiss Federal Statistical Office-Land Use Statistics (http://www.bfs.admin.ch/bfs/portal/de/index/themen/02/03/blank/data/gemeindedaten.html) 2009 data (German) accessed 25 March 2010

[7] http://www.hls-dhs-dss.ch/textes/d/D467.php

[8] http://www.hls-dhs-dss.ch/textes/f/F467.php

[9] http://www.hls-dhs-dss.ch/textes/i/I467.php

Grindelwald

Grindelwald	
 Grindelwald and Wetterhorn	
Country Switzerland **Canton** Bern **District** Interlaken-Oberhasli 	
Population	3796 (Dec 2011)[1]
- Density	22 /km^2 (57 /sq mi)
Area	km^2 (sq mi)[2]
Elevation	1034 m (ft)
Postal code	3818
SFOS number	0576
Mayor	Emanuel Schläppi
Localities	Alpiglen, Burglauenen, Grund, Itramen, Mühlebach, Schwendi, Tschingelberg, Wargistal
Surrounded by	Brienz, Brienzwiler, Fieschertal (VS), Guttannen, Innertkirchen, Iseltwald, Lauterbrunnen, Lütschental, Meiringen, Schattenhalb
Twin towns	Azumi, now Matsumoto (Japan)
Website	www.grindelwald.com [3] SFSO statistics [4]

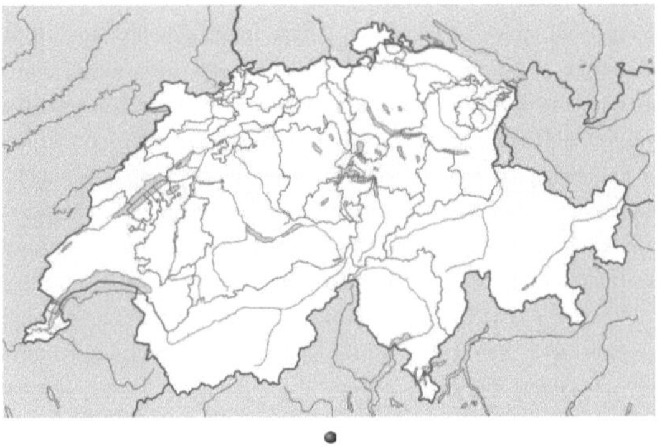

Grindelwald

Grindelwald is a village and municipality in the Interlaken-Oberhasli administrative district in the canton of Bern in Switzerland. Besides the village of Grindelwald, the municipality also includes the settlements of **Alpiglen**, **Burglauenen**, **Grund**, **Itramen**, **Mühlebach**, **Schwendi**, **Tschingelberg** and **Wargistal**.

Grindelwald village is located at 1034 m (ft) above sea level in the Bernese Alps.

History

Photochrom of Grindelwald between 1890 and 1905.

Grindelwald is first mentioned in 1146 as *Grindelwalt*.[4]

The oldest trace of a settlement in the area are scattered neolithic tools which have been discovered around Grindelwald village. Several Roman era coins have also been found in the municipality. A castle was built on the Burgbühl hill above the village during the High Middle Ages.[4]

In 1146 King Conrad III, granted estates in Grindelwald to Interlaken Abbey. In the late 12th century, the barons of the alpine valleys in what became the Berner Oberland went to war against the expansionist Duke Berthold V of Zähringen. The Duke defeated a coalition of nobles in the Grindlewald valley in 1191. His victory allowed him to expand Zähringen power into the Oberland, to expand the city of Thun and found the city of Bern. Beginning in the 13th century, Interlaken Abbey began to purchase rights and land in Grindelwald and eventually forced the local nobles out of the valley. The Abbey continued to exert influence in the village and in 1315 and again in 1332 ordered the villagers to raid Unterwalden to further the political ambitions of the Abbot's patrons. In response to the raids, in 1342, Unterwalden attacked Grindelwald. A few years later, in 1348-49, the villagers joined in an unsuccessful rebellion against ecclesiastical power. In 1528 the city of Bern adopted the Protestant Reformation and proceeded to spread the new faith in Grindelwald against the populations' wishes. Bern was able to impose its will, converted the village and secularized Interlaken Abbey and the Abbey's lands. Grindelwald became part of the bailiwick of Interlaken, under a Bernese bailiff.[4]

The first village church was a wooden building from the mid 12th century. The wooden building was replaced with the stone St. Mary's Church in 1180. This church was replaced in the 16th century and the present church was built in 1793.[4]

The tourism industry began in Grindelwald in the late 18th century as foreigners discovered the scenic town. Pictures of the vistas were widely reprinted, quickly making the village internationally famous. In the 19th century many Englishmen came to the village to climb the alpine peaks around the valley. The Finsteraarhorn (4274 m [Category:Pages with bad rounding precision ft]), the Wetterhorn (3692 m [Category:Pages with bad rounding precision ft]), the Eiger (3970 m [Category:Pages with bad rounding precision ft]), the Schreckhorn (4078 m [Category:Pages with bad rounding precision ft]) and the Gross Fiescherhorn (4049 m [Category:Pages with bad rounding precision ft]) were all climbed during the 19th century. The Grindelwald road was built in 1860-72 and the Bernese Oberland railway reached the village in 1890, both of which transformed an arduous journey into a simple trip and allowed tourists to flood into the village. The first resort opened in 1888 and there were 10 hotels in 1889 and by 1914 there were 33 in Grindelwald. A rack railway was built to Kleine Scheidegg in 1893 and it was expanded to the Jungfraujoch in 1912. Numerous ski lifts, cable cars, hiking trails and alpine huts were built in the late 19th and 20th centuries to allow tourists to explore the mountains. Today, almost the entire economy of Grindelwald is based on tourism.[4]

Geography

Grindelwald area from above

Grindelwald has an area of 171.08 km^2 (sq mi).[2] Of this area, 49.47 km^2 (sq mi) or 28.9% is used for agricultural purposes, while 28.02 km^2 (sq mi) or 16.4% is forested. Of the rest of the land, 3.06 km^2 (sq mi) or 1.8% is settled (buildings or roads), 1.37 km^2 (sq mi) or 0.8% is either rivers or lakes and 89.21 km^2 (sq mi) or 52.1% is unproductive land.[5]

Of the built up area, housing and buildings made up 1.0% and transportation infrastructure made up 0.6%. Out of the forested land, 12.9% of the total land area is heavily forested and 2.3% is covered with orchards or small clusters of trees. Of the agricultural land, 5.1% is pastures and 23.8% is used for alpine pastures. All the water in the municipality is flowing water. Of the unproductive areas, 6.6% is unproductive vegetation, 24.0% is too rocky for vegetation and 21.6% of the land is covered by glaciers.[5]

The municipality is quite large and is divided into seven mountain communities. However the municipality is dominated by the large tourist center of Grindelwald.

On 31 December 2009 Amtsbezirk Interlaken, the municipality's former district, was dissolved. On the following day, 1 January 2010, it joined the newly created Verwaltungskreis Interlaken-Oberhasli.[6]

Coat of arms

The blazon of the municipal coat of arms is *Gules on a Bar Argent a Semi Chamois rampant issuant Sable between seven, 4-3, Mullets of the second.*[7]

Demographics

Grindelwald has a population (as of December 2011) of 3796.[1] As of 2010, 18.0% of the population are resident foreign nationals.[8] Over the last 10 years (2000-2010) the population has changed at a rate of -1.3%. Migration accounted for 0.2%, while births and deaths accounted for -1.7%.[9]

Alphorn Festival in Grindelwald area

Most of the population (as of 2000) speaks German (3,531 or 86.8%) as their first language, Portuguese is the second most common (183 or 4.5%) and French is the third (69 or 1.7%). There are 60 people who speak Italian and 4 people who speak Romansh.[10]

New houses with buried garages in Grindelwald.

As of 2008, the population was 50.1% male and 49.9% female. The population was made up of 1,556 Swiss men (40.9% of the population) and 354 (9.3%) non-Swiss men. There were 1,568 Swiss women (41.2%) and 331 (8.7%) non-Swiss women.[8] Of the population in the municipality, 1,846 or about 45.4% were born in Grindelwald and lived there in 2000. There were 847 or 20.8% who were born in the same canton, while 510 or 12.5% were born somewhere else in Switzerland, and 679 or 16.7% were born outside of Switzerland.[10]

As of 2010, children and teenagers (0–19 years old) make up 17.5% of the population, while adults (20–64 years old) make up 62.5% and seniors (over 64 years old) make up 20%.[9]

As of 2000, there were 1,750 people who were single and never married in the municipality. There were 1,971 married individuals, 221 widows or widowers and 127 individuals who are divorced.[10]

As of 2000, there were 679 households that consist of only one person and 82 households with five or more people. In 2000, a total of 1,581 apartments (44.9% of the total) were permanently occupied, while 1,750 apartments (49.6%) were seasonally occupied and 194 apartments (5.5%) were empty.[11] As of 2010, the construction rate of new housing units was 13.1 new units per 1000 residents.[9]

The historical population is given in the following chart:[4] [12]

Heritage sites of national significance

The Jungfraubahn, a mountain railway up the Jungfrau mountain, is listed as a Swiss heritage site of national significance. The land around the Kleine Scheidegg is part of the Inventory of Swiss Heritage Sites.[13]

Jungfraubahn Mountain Railway

Kleine Scheidegg

Politics

In the 2011 federal election the most popular party was the Swiss People's Party (SVP) which received 54.2% of the vote. The next three most popular parties were the Conservative Democratic Party (BDP) (14.8%), the FDP.The Liberals (8.8%) and the Green Party (6.9%). In the federal election, a total of 1,194 votes were cast, and the voter turnout was 44.4%.[14]

Economy

Hotels in Grindelwald. The local economy is mostly dependent on tourism.

As of 2011, Grindelwald had an unemployment rate of 1.48%. As of 2008, there were a total of 2,714 people employed in the municipality. Of these, there were 334 people employed in the primary economic sector and about 130 businesses involved in this sector. 393 people were employed in the secondary sector and there were 51 businesses in this sector. 1,987 people were employed in the tertiary sector, with 229 businesses in this sector.[9] There were 2,403 residents of the municipality who were employed in some capacity, of which females made up 45.4% of the workforce.

In 2008 there were a total of 2,265 full-time equivalent jobs. The number of jobs in the primary sector was 165, of which 159 were in agriculture and 7 were in forestry or lumber production. The number of jobs in the secondary sector was 365 of which 42 or (11.5%) were in manufacturing, 6 or (1.6%) were in mining and 292 (80.0%) were in construction. The number of jobs in the tertiary sector was 1,735. In the tertiary sector; 241 or 13.9% were in wholesale or retail sales or the repair of motor vehicles, 193 or 11.1% were in the movement and storage of goods, 1,022 or 58.9% were in a hotel or restaurant, 27 or 1.6% were the insurance or financial industry, 45 or 2.6% were technical professionals or scientists, 46 or 2.7% were in education and 58 or 3.3% were in health care.[15]

In 2000, there were 432 workers who commuted into the municipality and 265 workers who commuted away. The municipality is a net importer of workers, with about 1.6 workers entering the municipality for every one leaving.[16] Of the working population, 9.2% used public transportation to get to work, and 36.2% used a private car.[9]

Religion

From the 2000 census, 690 or 17.0% were Roman Catholic, while 2,874 or 70.6% belonged to the Swiss Reformed Church. Of the rest of the population, there were 31 members of an Orthodox church (or about 0.76% of the population), there were 2 individuals (or about 0.05% of the population) who belonged to the Christian Catholic Church, and there were 105 individuals (or about 2.58% of the population) who belonged to another Christian church. There was 1 individual who was Jewish, and 32 (or about 0.79% of the population) who were Islamic. There were 14 individuals who were Buddhist and 1 individual who belonged to another church. 144 (or about 3.54% of the population) belonged to no church, are agnostic or atheist, and 227 individuals (or about 5.58% of the population) did not answer the question.[10]

Climate

Between 1981 and 2010 Grindelwald had an average of 145.4 days of rain or snow per year and on average received 1450 mm (in) of precipitation. The wettest month was August during which time Grindelwald received an average of 165 mm (in) of rain or snow. During this month there was precipitation for an average of 14.4 days. The month with the most days of precipitation was June, with an average of 14.8, but with only 151 mm (in) of rain or snow. The driest month of the year was February with an average of 89 mm (in) of precipitation over 9.7 days.[17]

Transport

Grindelwald railway station, in the centre of Grindelwald village, is served by trains of the Berner Oberland Bahn (BOB), which run to Interlaken, and by trains of the Wengernalpbahn (WAB), which run to Kleine Scheidegg and on to Wengen and Lauterbrunnen. At Kleine Scheidegg connection can be made with the Jungfraubahn, which ascends inside the Eiger to the Jungfraujoch.[18]

Grindelwald railway station

Besides Grindelwald station, there are five other railway stations within the municipality of Grindelwald. These are Burglauenen and Schwendi on the BOB, and Grindelwald Grund, Brandegg and Alpiglen on the WAB.[18]

The Gondelbahn Grindelwald-Männlichen connects Grindelwald with the Männlichen and with onward travel on the Luftseilbahn Wengen-Männlichen offers an alternative route to Wengen.

The minor summit of First is accessible by ski lift from Grindelwald.

Winter sports

Long famed as a winter tourist destination with slopes for beginners, intermediates and the challenges of the Eiger glacier for the experienced, there are activities for the non-skiers, from tobogganing to groomed winter hiking tracks. It is the usual starting point for ascents of the Eiger and the Wetterhorn. Nowadays Grindelwald is also a popular summer activity resort with many miles of hiking trails across the Alps.

In the media

Many scenes of the documentary film *The Alps* were shot in the region of Grindelwald, particularly on the north face of the Eiger. The James Bond film *On Her Majesty's Secret Service* includes a chase through a skating rink and Christmas festival in Grindelwald.[19] Grindelwald's mountains were used as the basis for the view of Alderaan in *Star Wars Episode III: Revenge of the Sith*.[20]

The Clint Eastwood film "The Eiger Sanction" was shot in Grindelwald and Kleine Scheidegg.

Some of the action scenes in *The Golden Compass* were also shot in Grindelwald.[21]

Grindelwald appears as a race course in Gran Turismo 2.

Grindelwald is also used as a character name for Gellert Grindelwald in the *Harry Potter* series.

In the eight leg of "The amazing race 3" season three, a roadblock destination and a pit stop was located in Grindelwald. In "The amazing race 22", (also leg eight) Grindelwald were the location of a roadblock, a switchback and a pitstop.

Education

In Grindelwald about 1,647 or (40.5%) of the population have completed non-mandatory upper secondary education, and 302 or (7.4%) have completed additional higher education (either university or a *Fachhochschule*). Of the 302 who completed tertiary schooling, 57.9% were Swiss men, 24.5% were Swiss women, 9.3% were non-Swiss men and 8.3% were non-Swiss women.[10]

The Canton of Bern school system provides one year of non-obligatory Kindergarten, followed by six years of Primary school. This is followed by three years of obligatory lower Secondary school where the students are separated according to ability and aptitude. Following the lower Secondary students may attend additional schooling or they may enter an apprenticeship.[22]

During the 2010-11 school year, there were a total of 341 students attending classes in Grindelwald. There were 3 kindergarten classes with a total of 55 students in the municipality. Of the kindergarten students, 27.3% were permanent or temporary residents of Switzerland (not citizens) and 20.0% have a different mother language than the classroom language. The municipality had 12 primary classes and 187 students. Of the primary students, 15.0% were permanent or temporary residents of Switzerland (not citizens) and 12.8% have a different mother language than the classroom language. During the same year, there were 6 lower secondary classes with a total of 99 students. There were 8.1% who were permanent or temporary residents of Switzerland (not citizens) and 5.1% have a different mother language than the classroom language.[23]

As of 2000, there were 20 students in Grindelwald who came from another municipality, while 39 residents attended schools outside the municipality.[16]

Grindelwald is home to the *Bibliothek Grindelwald* library. The library has (as of 2008) 6,144 books or other media, and loaned out 10,777 items in the same year. It was open a total of 136 days with average of 6.5 hours per week during that year.[24]

Famous residents

- Richard Wagner
- Martina Schild, alpine skier, runner-up in the 2006 Winter Olympics women's downhill race
- Hedy Schlunegger, Olympic champion 1948 in downhill skiing
- Oleg Protopopov and Ludmila Belousova, 1964 and 1968 Olympic figure skating champions

See also

- Swiss Alps

References

[1] Swiss Federal Statistical Office - STAT-TAB (http://www.pxweb.bfs.admin.ch/Database/German_01 - BevÃ¶lkerung/01.2 - BevÃ¶lkerungsstand und -bewegung/01.2 - BevÃ¶lkerungsstand und -bewegung.asp?lang=1&prod=01&secprod=2&openChild=true), online database – *Datenwürfel für Thema 01.2 - Bevölkerungsstand und -bewegung* **(German)** accessed 3 October 2012

[2] Arealstatistik Standard - Gemeindedaten nach 4 Hauptbereichen (http://www.bfs.admin.ch/bfs/portal/de/index/themen/02/03/blank/data/gemeindedaten.html)

[3] http://www.grindelwald.com

[4] *Grindelwald* in German (http://www.hls-dhs-dss.ch/textes/d/D331.php), French (http://www.hls-dhs-dss.ch/textes/f/F331.php) and Italian (http://www.hls-dhs-dss.ch/textes/i/I331.php) in the online *Historical Dictionary of Switzerland*.

[5] Swiss Federal Statistical Office-Land Use Statistics (http://www.bfs.admin.ch/bfs/portal/de/index/themen/02/03/blank/data/gemeindedaten.html) 2009 data **(German)** accessed 25 March 2010

[6] Nomenklaturen – Amtliches Gemeindeverzeichnis der Schweiz (http://www.bfs.admin.ch/bfs/portal/de/index/infothek/nomenklaturen/blank/blank/gem_liste/02.html) **(German)** accessed 4 April 2011

[7] Flags of the World.com (http://www.crwflags.com/fotw/flags/ch-be160.html) accessed 18-March-2013

[8] Statistical office of the Canton of Bern (http://www.fin.be.ch/de/index/finanzen/finanzen/publikationen/wohnbevoelkerung.html) **(German)** accessed 4 January 2012

[9] Swiss Federal Statistical Office (http://www.bfs.admin.ch/bfs/portal/en/index/regionen/02/key.html) accessed 18 March 2013

[10] STAT-TAB Datenwürfel für Thema 40.3 - 2000 (http://www.pxweb.bfs.admin.ch/Database/German_40 - EidgenÃ¶ssische VolkszÃ¤hlung/40.3 - 2000/40.3 - 2000.asp?lang=1&prod=40&secprod=3&openChild=true) **(German)** accessed 2 February 2011

[11] Swiss Federal Statistical Office STAT-TAB - Datenwürfel für Thema 09.2 - Gebäude und Wohnungen (http://www.pxweb.bfs.admin.ch/Database/German_09 - Bau- und Wohnungswesen/09.2 - GebÃ¤ude und Wohnungen/09.2 - GebÃ¤ude und Wohnungen.asp?lang=1&prod=09&secprod=2&openChild=true) **(German)** accessed 28 January 2011

[12] Swiss Federal Statistical Office STAT-TAB Bevölkerungsentwicklung nach Region, 1850-2000 (http://www.pxweb.bfs.admin.ch/Database/German_40 - EidgenÃ¶ssische VolkszÃ¤hlung/40.4 - 1850-2000/40.4 - 1850-2000.asp?lang=1&prod=40&secprod=4&openChild=true) **(German)** accessed 29 January 2011

[14] Swiss Federal Statistical Office 2011 Election (http://www.bfs.admin.ch/bfs/portal/de/index/themen/17/02/blank/data/05/03.html) **(German)** accessed 8 May 2012

[15] Swiss Federal Statistical Office STAT-TAB Betriebszählung: Arbeitsstätten nach Gemeinde und NOGA 2008 (Abschnitte), Sektoren 1-3 (http://www.pxweb.bfs.admin.ch/Database/German_06 - Industrie und Dienstleistungen/06.2 - Unternehmen/06.2 - Unternehmen.asp?lang=1&prod=06&secprod=2&openChild=true) **(German)** accessed 28 January 2011

[16] Swiss Federal Statistical Office - Statweb (http://www.media-stat.admin.ch/stat/pendler/pop.php) **(German)** accessed 24 June 2010

[17] , the Grindelwald weather station elevation is 1158 meters above sea level.

[18] Swiss Confederation. *map.geo.admin.ch* (http://map.geo.admin.ch/?selectedNode=node_ch.swisstopo.swissboundaries3d-kanton-flaeche.fill1&Y=645508&X=163811&zoom=8&bgLayer=ch.swisstopo.pixelkarte-farbe&layers=ch.swisstopo.swissboundaries3d-gemeinde-flaeche.fill&layers_opacity=1&layers_visibility=true&lang=en) (Map). . Retrieved 2013-01-28.

[19] IMDB-On Her Majesty's Secret Service (http://www.imdb.com/title/tt0064757/)

[20] IMDB-Star Wars Episode III: Revenge of the Sith (http://www.imdb.com/title/tt0121766/locations)

[21] IMDB-The Golden Compass (http://www.imdb.com/title/tt0385752/locations)

[22] EDK/CDIP/IDES (2010). Kantonale Schulstrukturen in der Schweiz und im Fürstentum Liechtenstein / Structures Scolaires Cantonales en Suisse et Dans la Principauté du Liechtenstein (http://edudoc.ch/record/35128/files/Schulsystem_alle.pdf) (Report). . Retrieved 24 June 2010.

[23] Schuljahr 2010/11 pdf document (http://www.erz.be.ch/erz/de/index/direktion/organisation/generalsekretariat/statistik/schul-_und_gemeindetabellen.html)**(German)** accessed 4 January 2012

[24] Swiss Federal Statistical Office, list of libraries (http://www.bfs.admin.ch/bfs/portal/de/index/themen/16/02/02/data.html) **(German)** accessed 14 May 2010

External links

- Grindelwald (http://www.grindelwald.com) official website
- Webcams (http://www.swisspanorama.com/)
- *Grindelwald* in German (http://www.hls-dhs-dss.ch/textes/d/D331.php), French (http://www.hls-dhs-dss.ch/textes/f/F331.php) and Italian (http://www.hls-dhs-dss.ch/textes/i/I331.php) in the online *Historical Dictionary of Switzerland*.

Engelberg

Engelberg

Country	Switzerland
Canton	Obwalden
District	n.a.

Population	3939 (Dec 2011)[1]
- Density	97 /km^2 (250 /sq mi)
Area	km^2 (sq mi)[2]
Elevation	1000 m (ft) Engelberg town
• Highest	3238 m - Mount Titlis
• Lowest	560 m - Grafenort
Postal code	6390
SFOS number	1402
Surrounded by	Attinghausen (UR), Gadmen (BE), Innertkirchen (BE), Isenthal (UR), Wassen (UR), Wolfenschiessen (NW)
Website	www.gde-engelberg.ch [3] SFSO statistics [4]

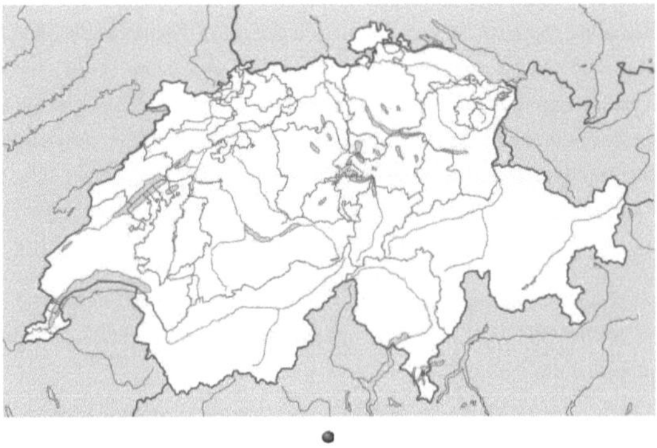

Engelberg

Engelberg is a resort town and municipality in the canton of Obwalden in Switzerland. Besides the town of Engelberg, the municipality also includes the settlements of **Grafenort**, **Obermatt** and **Schwand**.

The municipality of Engelberg is an exclave of Obwalden, surrounded by the cantons of Bern, Nidwalden and Uri. Engelberg has a population (as of 31 December 2011) of 3939.[1] As of 2007, 21.0% of the population was made up of foreign nationals.[4]

Hubert Sattler (1817–1904): View of Engelberg

Engelberg is the leading mountain resort (Urner Alps) in central Switzerland. In the Middle Ages, Engelberg was known for the educational accomplishments of its Benedictine monastery, the Engelberg Abbey, whose school was well-known and regarded throughout the country. From the 19th Century, Engelberg became internationally known as a resort and spa, but is today visited as much for skiing as for its clean air. With its combination of modern sports facilities and alpine location, Engelberg is a magnet for both summer and winter tourism. The closest large cities are Lucerne and Zurich.

View of Engelberg from Hotel Ramada Balcony

History

Engelberg is first mentioned as *Engilperc* in 1122, when the Abbey was first founded there, although the mountain pasture of Trübsee was already exploited collectively before this time.

From 1850, Engelberg became an international vacation resort (mineral water, milk serum and fresh air cures). Many hotels were built by the families Cattani, Hess and Odermatt, pioneers of tourism. From 1872 to 1874, a new, wider road was built, and the Stansstad-Engelberg electric railway was opened in 1898.[5]

Engelberg-Titlis cableway

Hiking and other mountain sports developed at the end of 19th Century and Engelberg first held a winter season in 1903-1904. Since 1913, a funicular railway connects Engelberg to Gerschnialp and, from there, the second cable car in Switzerland (1927) led to Trübsee. The decade preceding the First World War was a period of boom conditions (165,922 visitor-nights in 1911). The widening of the road and the extension of the railway to Luzern (1964) considerably opened up the tourism catchment area of the station and, in 1967, the higher section of the Titlis cable car was opened. Recently, regular conferences in Engelberg came to supplement winter tourism. In 2000, the tertiary sector, especially tourism, offered three quarters of the employment of Engelberg.[5]

Geography

Engelberg has an area of 74.8 square kilometers (sq mi). Of this area, 28.5% is used for agricultural purposes, while 24.5% is forested. Of the rest of the land, 3.1% is settled (buildings or roads) and the remainder (43.9%) is non-productive (rivers, glaciers or mountains).[6]

The average altitude of Engelberg is 1020 m (ft). However the town is surrounded by the Alps, creating very steep terrain. The highest point in the borders of the town is Mt. Titlis which is 3239 m (ft).

Transport

The municipality of Engelberg is served by two stations on the Luzern–Stans–Engelberg line. Engelberg station is located within the resort of Engelberg and is the terminus of the line. Grafenort station lies to the north, one station down the line. Both stations are served by hourly InterRegio trains from the city of Lucerne.[7]

Tourism

The ski terrain is largely for advanced skiers, and although there are some easy places to ski, beginners should be wary of some challenging pistes. It is a snow-sure ski resort, with an unusually long season, generally opening for skiing at the start of October and remaining open until the end of May.

It's also a major summer resort, with many activities available both at village level and on the mountain. The lift system operates throughout the year, allowing a wide range of walking and cycling trails to be available even to the more fitness-challenged visitor.

In recent years, Engelberg has become something of an off-piste mecca, with many winter visitors, particularly from Scandinavia, coming specifically for its many challenging descents. Many of these are accessible from the lift-served ski areas of Titlis or Joch Pass, on the predominantly North-facing slopes leading down from Mt Titlis.

On the opposite side of the valley, the ski area of Brunni is much more suitable for beginners and early intermediates, but being South-facing it doesn't have quite the same snow reliability and length of season.

Engelberg also hosts a round of the ski jumping world cup at Gross-Titlis-Schanze.

The main sights of Engelberg are: Mount Titlis, Mount "Hahnen", Engelberg Abbey, a Benedictine monastery in the Swiss Congregation, the museum of the valley, the hotel *Europäischer Hof*, a number of old chapels.

Scenes from many Bollywood films are shot at Engelberg, and the area is consequently very popular with Indian tourists.

Gross-Titlis-Schanze

Demographics

Engelberg has a population (as of 31 December 2011) of 3939.[1] As of 2007, 21.0% of the population was made up of foreign nationals.[4] Over the last 10 years the population has grown at a rate of 7.1%. Most of the population (as of 2000) speaks German as their mother tongue (88.2%), with Serbo-Croatian being second most common (2.5%) and English being third (2.2%).[6] As of 2000 the gender distribution of the population was 49.9% male and 50.1% female. As of 2000 there are 1,650 households in Engelberg.[4]

In the 2007 election the most popular party was the SVP which received 37.4% of the vote. The next three most popular parties were a variety of other parties (not major) (28.6%), the CVP (22.7%) and the SPS (11.3%).

The age distribution of the population (as of 2000) is children and teenagers (0–19 years old) make up 26.2% of the population, while adults (20–64 years old) make up 58.3% and seniors (over 64 years old) make up 15.4%. In Engelberg about 65.5% of the population (between age 25-64) have completed either non-mandatory upper secondary education or additional higher education (either university or a *Fachhochschule*).[6]

The historical population is given in the following table:[5]

year	population
1709	678
1799	1,468
1850	1,737
1900	1,973
1950	2,544
2000	3,544

Business and industry

There are 320 local businesses which employ 1700 people. 11% of these are in the agricultural sector, 14% in trade and industry, 75% in services.

Engelberg has an unemployment rate of 1.22%. As of 2005, there were 176 people employed in the primary economic sector and about 65 businesses involved in this sector. 227 people are employed in the secondary sector and there are 36 businesses in this sector. 1,295 people are employed in the tertiary sector, with 186 businesses in this sector.[6]

Due to the risks of filming in the disputed region of Kashmir, many Bollywood films requiring a Kashmir/snowy mountain setting are filmed in Engelberg.

Cultural references

Engelberg is mentioned in chapter XXIV of Henry James's 1875 novel *Roderick Hudson.*

Giuseppe Piazzi bishop of Bergamo died in Engelberg in 1963.

Climate

Between 1961 and 1990 Engelberg had an average of 157.9 days of rain per year and on average received 1510 mm (in) of precipitation. The wettest month was August during which time Engelberg received an average of 185 mm (in) of precipitation. During this month there was precipitation for an average of 14.7 days. The month with the most days of precipitation was June, with an average of 16.1, but with only 179 mm (in) of precipitation. The driest month of the year was February with an average of 90 mm (in) of precipitation over 14.7 days.[8]

References

[1] Swiss Federal Statistics Office – STAT-TAB (http://www.pxweb.bfs.admin.ch/Dialog/varval.asp?ma=px-d-01-2A01&ti=Ständige+und+Nichtständige+Wohnbevölkerung+nach+Region,+Geschlecht,+Nationalität+und++Alter&path=../Database/German_01 - Bevölkerung/01.2 - Bevölkerungsstand und -bewegung/&lang=1&prod=01&openChild=true&secprod=2) *Ständige und Nichtständige Wohnbevölkerung nach Region, Geschlecht, Nationalität und Alter* (German) accessed 3 October 2012

[2] Arealstatistik Standard - Gemeindedaten nach 4 Hauptbereichen (http://www.bfs.admin.ch/bfs/portal/de/index/themen/02/03/blank/data/gemeindedaten.html)

[3] http://www.gde-engelberg.ch

[4] Canton Obwalden Statistics (http://www.statistik-obwalden.ch/index.cfm?navid=78&docid=78) accessed 2 September 2009 (German)

[5] *Engelberg* in German (http://www.hls-dhs-dss.ch/textes/d/D741.php), French (http://www.hls-dhs-dss.ch/textes/f/F741.php) and Italian (http://www.hls-dhs-dss.ch/textes/i/I741.php) in the online *Historical Dictionary of Switzerland.*

[6] Swiss Federal Statistical Office (http://www.bfs.admin.ch/bfs/portal/en/index/regionen/02/key.html) accessed 03-Sep-2009

[8] , the weather station elevation is 1035 meters above sea level.

[9] The weather station elevation is 1,035 meters above sea level.

External links

- Engelberg Tourism (http://www.engelberg.ch)
- Free Pictures Engelberg (http://www.pictures-switzerland.com/engelberg/)
- Titlis Glacier (http://www.titlis.ch)
- Ski Club of Great Britain Resort Info (http://www.skiclub.co.uk/skiclub/skiresorts/resortintro.aspx/Engelberg)
- Brunni area (http://www.brunni.ch/)
- *Engelberg* in German (http://www.hls-dhs-dss.ch/textes/d/D741.php), French (http://www.hls-dhs-dss.ch/textes/f/F741.php) and Italian (http://www.hls-dhs-dss.ch/textes/i/I741.php) in the online *Historical Dictionary of Switzerland.*

Article Sources and Contributors

Innertkirchen *Source*: http://en.wikipedia.org/w/index.php?title=Innertkirchen *Contributors*: BillC, Chatfecter, Chris j wood, D6, Dbachmann, DidiWeidmann, MadGeographer, Samtheboy, Tobyc75, 4 anonymous edits

Communes_of_Switzerland *Source*: http://en.wikipedia.org/w/index.php?title=Communes_of_Switzerland *Contributors*: 159753, 2Neuenburg, Acntx, Amikake3, Andonic, Booksworm, Chanheigeorge, Corti, Crusoe8181, Cusio, D6, Db1944, Dbachmann, Dirrival, Docu, Elipongo, Emerson7, EnOreg, Felix Portier, Frietjes, Geschichte, Gestumblindi, Grutness, Hadal, Haruo, Irmgard, J. 'mach' wust, Jurema Oliveira, LeoNomis, Lugia2453, MBisanz, MK, Megan1967, Melo86, NawlinWiki, Onco p53, PsY.cHo, Reindra, Rich Farmbrough, Roland zh, RubleuleR, Schutz, Sectryan, Sietse Snel, Srleffler, Tobias Conradi, Tobyc75, Ultimate Destiny, Wmahan, Woohookitty, Yellowdesk, 27 anonymous edits

Interlaken-Oberhasli_(administrative_district) *Source*: http://en.wikipedia.org/w/index.php?title=Interlaken-Oberhasli_%28administrative_district%29 *Contributors*: Chris j wood, Dbachmann, DidiWeidmann, Schwyz, Tobyc75

Cantons_of_Switzerland *Source*: http://en.wikipedia.org/w/index.php?title=Cantons_of_Switzerland *Contributors*: 05, 159753, 2Neuenburg, 3riol, A bit iffy, A. Parrot, AP1787, Acntx, Ahuskay, Alarob, Aram33, Argasp, Arthena, Bag, Barticus88, Bearcat, Blanche of King's Lynn, Brammen, Briséis, Brittainia, BrownHairedGirl, CUSENZA Mario, Cassowary, Chanheigeorge, Chris j wood, ChrisfromHouston, Chùn-hiàn, Claude girardin, Closedmouth, Cnilep, Cunado19, D6, DDima, Dbachmann, Dcoetzee, Deacon of Pndapetzim, Denelson83, Djinn112, Docu, Dzordzm, EamonnPKeane, EdwinHJ, El C, Electionworld, Enlil Ninlil, Ericamick, Fan-1967, Filzstift, Frank101, Gazilion, Gestumblindi, Gurch, Haddiscoe, Hashar, Hayden120, Hinote, Hmains, Hroðulf, Hunnjazal, Hvn0413, Indigoeleven, Insilvis, Janburse, JdeJ, Jschwa1, Jusjih, Kahusi, Kerryh r, Kokiri, Korovioff, Lightmouse, Lkmorlan, Longbow4u, Luuva, MadGeographer, Martg76, Mascia Luccova, Master of Puppets, Math Champion, Maxí, Megan1967, MeltBanana, Mistakefinder, MrRadioGuy, Mrt3366, Mschlindwein, Mthwaite, Murtasa, N2e, Nalta, Ngagnebin, Nilfanion, Nohat, Nowhither, Ojw, OwenBlacker, Pascal, Patrick, Paul-L, Piccadilly, Piccolo Modificatore Laborioso, Pne, Postdlf, PsY.cHo, Random user 39849958, Rich Farmbrough, Rjwilmsi, Roke, Rpckvv, Sallygirl1783, Sam Blacketer, Sandstein, Sardanaphalus, Schutz, ScottMainwaring, Sesel, SetRochele, SetRochelewsad, Sherool, SiBr4, Skew-t, Sluzzelin, Smjg, Stephensuleeman, Stevietheman, Tamfang, TenOfAllTrades, Thorsen, Tobias Conradi, Tobyc75, Tonymec, Ultimate Destiny, Uncle Scrooge, Vdiest, West.andrew.g, Woohookitty, Worobiew, Wshun, Zacharie Grossen, Zondor, Žiga, პაატა შ, 103 anonymous edits

Canton_of_Bern *Source*: http://en.wikipedia.org/w/index.php?title=Canton_of_Bern *Contributors*: 2Neuenburg, 93foye, AP1787, Adlei86, Ahoerstemeier, Ajnem, Al Silonov, Angr, Attilios, Barewires, Bastin, Bejnar, Benjiwolf, Bongoramsey, CarolGray, CdaMVvWgS, Chris j wood, Claude girardin, Codrinb, Coelacan, CommonsDelinker, Cú Faoil, D6, David Newton, Dbachmann, Deacon of Pndapetzim, Denelson83, Docu, Dumbassguinea, Ellywa, Ericoides, Feldspatz, Funnyhat, Gertjan R., Goldfishbutt, Goonies, Grafen, Habib87, Hayden120, Henning Makholm, Hunnjazal, Hvn0413, Ihorm1968, J. 'mach' wust, John of Reading, Johnbod, Joseph Solis in Australia, Jpgordon, Katalaveno, Khazar2, KirkEN, Kokiri, Kwamikagami, Kyng, Leavit2stever, LightPhoenix, Longbow4u, Lupo, MIKHEIL, MadGeographer, Magioladitis, Marquez, Mazca, MeltBanana, Mulhim, Neuenburg, Ngagnebin, Nightworker, Nk, Numbo3, ONUnicorn, Ohconfucius, OwenBlacker, PaulTheDon, Pedant17, Perhelion, PhnomPencil, Piccolo Modificatore Laborioso, Pmanderson, Poolback, Quadpus, RichF, Rjwilmsi, S. Neuman, Saltose, Sandstein, Sasquatch, Schutz, Schwyz, Sleigh, Snowdog, Squash Racket, Swissjane, Tabletop, TheSwitzerdude, Tobyc75, Tombomp, Tooto, Trilobite, Trumpnewtq, Unyoyega, Uuakamo, Van der Hoorn, Voyager, Walden69, WereSpielChequers, WhisperToMe, Wik, Wikibob, Wonder al, Writtenright, Zacharie Grossen, Zumbo, Žiga, 91 anonymous edits

Coat_of_arms *Source*: http://en.wikipedia.org/w/index.php?title=Coat_of_arms *Contributors*: 123aidan, 1exec1, 28421u2232nfenfcenc, A More Perfect Onion, A little insignificant, A.zstephens, ABF, ATurtle05, Addshore, Adhyatma3010, Aervanath, Aetylus, Agentscott00, Ahoerstemeier, Ahuskay, Aka042, Alansohn, Alexius08, Amateur55, Amjad Abdullah, Andre Engels, Angela, AnonMoos, Anonymous from the 21st century, Antalagy, Apeloverage, Arab Hafez, Araignee, ArnoldReinhold, Artyom, Arwel Parry, Asbestos, Ashish Bakshi, Astrotrain, Attilios, BIL, Bakabaka, BanyanTree, Barneca, Basil Bernar, Beetstra, Bennylin, Betacommand, Bewareofdog, Big Adamsky, BilCat, Bissinger, Blauendorn, Bobo192, BolinhasFofas, Borameer, Brandon5485, Brian0918, Brianga, BryanG, Bubbha, C9Rav9c, CALR, CTZMSC3, Cahk, CambridgeBayWeather, Can't sleep, clown will eat me, Capricorn42, CardinalDan, Care, CaribDigita, Carlsmith, Celtus, Centauri, Charruyego, Checkpoint, ChrisOelmueller, Chrism, Christian75, Chwyatt, Circeus, Clear&Simple, Codex Sinaiticus, CommonsDelinker, Cordless Larry, Correogsk, Craigy144, CristianChirita, Crzrussian, Curb Chain, Custodiansoftime, Czar Brodie, DARTH SIDIOUS 2, DTRY, Danbarnesdavies, Danceswithzerglings, Dancingwombatsrule, Danrok, Darcij, Darth Panda, David Latapie, David Shay, David de Cooman, Dawn Bard, Dbachmann, Dcandeto, Debouncer, Decora, Delirium, Demiurge, Derek Andrews, DeusMP, Dfflick319, Digger845, Discospinster, Dlae, Dogears, Doops, Dr pda, DrHollisCollier, Dthomsen8, Duffman, Dysepsion, ESkog, Eliz81, Ellmist, EncycloPetey, Enviroboy, Enzo Aquarius, Epbr123, Equendil, Eric-Wester, Euchiasmus, Eulyzes, Evadb, Evertype, Everyking, Evil saltine, Evrik, Ezeu, Fanatix, Fengxianjin, FiggyBee, Finavon, Fiyinite7, Flausy, Foxcommunication, FrancoGG, Freakofnurture, FrederickSommer, Freederick, Fritzpoll, Fvasconcellos, Gakusha, Galoubet, Galwhaa, Garryq, GearedBull, Generalseven, Geniac, Geoffrey.landis, George.moug, Gilliam, Gimboid13, Gimmetrow, Gire 3pich2005, Glane23, Glenlarson, Gogo Dodo, Gorovich, Gregmatthews, Group29, Grutness, Gsp8181, Gtstricky, Gustav von Humpelschmumpel, Gustavo Szwedowski de Korwin, Hadžija, Halmstad, HamburgerRadio, Hawjam, Heraldics, Herostratus, Hindustanilanguage, Hmains, Hoss789, Hu12, Huntscorpio, INkubusse, IRP, Iaberis, Iamunknown, Igoldste, Imagecrafting, ImperatorExercitus, Inka 888, Innerche, Ipankonin, Iridescent, Islescape, J.delanoy, J36miles, JFreeman, Jackol, JackyR, Jake Wartenberg, JamesBWatson, Jaykjellybean, Jdforrester, JerryFriedman, Jhamez84, Jiang, Jjojj, Jncraton, John, John254, JohnCD, Johnathan Tsui, Jonjames1986, Joseph Solis in Australia, José Luís Ávila Silveira, Jschnur, Jusdafax, Jyusin, Kahner5150, Kaihsu, Kanguole, Kansoku, Karmosin, Keilana, Kelisi, Kevin96850, Khoikhoi, Kimon, King of Hearts, KingNewbs, Kintetsubuffalo, Kiore, Kittybrewster, Kookyunii, Kool Lat'n SD, Kostisl, Kovacho, Kpgh99, Krylonblue83, Kylu, Kyuko, Lacrimosus, Lahiru k, Lambiam, Lammoth, Larry1972, LeaveSleaves, LiDaobing, LibLord, Linkboyz, Lobsterthermidor, Logan, Lord Brantus, Lord Fitzwarin, Loren.wilton, Lradrama, LtNOWIS, M.K, M0rphzone, MER-C, MacGonagall, Madhero88, Magicuser3, Magioladitis, Magog the Ogre, Man vyi, Marcelivan, Mark, Martykoch, MathHisSci, Mauls, Maycoll F. Vieira, Mb1000, Mellisa Anthony Jones, Mhking, Mich Taylor, Michael Hardy, Michaelsanders, Midgrid, Mike Rosoft, Milford212, Mingwangx, Mon Vier, Mormegil, Mothmolevna, Mouser, Moyogo, Munboy, Muriel Gottrop, Mushroom, MusikAnimal, Mygerardromance, Nadav1, Nakon, Nameneko, Neurolysis, NewEnglandYankee, Nilfanion, Nivix, Nkcs, Noropdoropi, Nsaa, Ntsimp, OHCLOUDSUNFOLD, Octane, Ohconfucius, Orange Suede Sofa, Owned Souls, Oxymoron83, PBS, PDH, Palnatoke, Paul Barlow, Paul venter, Pdcook, Pe-Jo, Persian Poet Gal, Petri Krohn, PhJ, Phaedriel, Philip Trueman, Philipandrew, Pieneman1, Pinethicket, Pinkadelica, Pkhg199, PleaseStand, Pmanderson, Pne, Poco a poco, Pojanji, Pokrajac, Porqin, Possum, Prothonotar, Puddhe, Quite, Qxz, Qyd, RandomStringOfCharacters, RapidR, Razakel19, Rdsmith4, Remuel, Rettetast, Reywas92, Rhurtadon, Ricardocolombia, Richard A McDonald, RichardGonzalesV, Rjstott, Robchurch, Rockhall, Ronhjones, Roux, Royboycrashfan, S, SChockey97, SPhotographer, Sadalmelik, Saga City, Salvadorjo, Samuella99, Sbwoodside, Schockey14, SchreiberBike, SeanR70, Sephiroth storm, Shanel, Silverhorse, SimonP, Simonmetcalf, Sinaloa, Sinhaempire, Sionus, Siroxo, Sistan, Skater, Sketraw, Skull 'n' Femurs, Sky Harbor, SmallPotatoes, Smalljim, Smoon17, Snow Blizzard, Snowmanradio, Snowolf, Sotakeit, SowelBlack, Spencer, Sputnikcccp, Srnec, Stansult, Strike Eagle, Studioce, SupermanReturns, SusikMkr, Sweetloulou, TJ Spyke, Tabletop, Tamfang, Tauex, Tckma, Techman224, Tedder, Teo64x, Terot, Terra Xin, TerraHikaru, Thakurji, The Thing That Should Not Be, Theone00, Thingg, This user has left wikipedia, Thisbites, Thisisborin9, Thomas Blomberg, Thumperward, ThuranX, Tide rolls, Tilstock, Tirthajyoti, Tmopkisn, Tom.k, Tone, Tozzano, Trusilver, Twinxor, TylerDurden1963, Tyrenius, Uhai, UserDoe, Utcursch, VAwebteam, Valentinian, Vanished User 1004, VanishedUserABC, Vasilcho, Vincent Liu, Vrenator, Vuo, Vzbs34, WJBscribe, Wales, Wayne Slam, Whistlesgowhoo, Whpq, Wikipedian2, Wilhelm meis, Witan, Woohookitty, Writerguy, Writtenright, Xanderliptak, Xn4, Yamakiri, Yarp, Yopie, Yworo, Zalgo, Zalhael, Zanhe, Zarthon, Zaxo1994, , , 958 anonymous edits

Agricultural_cooperative *Source*: http://en.wikipedia.org/w/index.php?title=Agricultural_cooperative *Contributors*: APB-CMX, Agricmarketing, Alan Liefting, Altenmann, AmishThrasher, Anandbharti, Annak123, Anthony Appleyard, BD2412, Baseball Bugs, Bejnar, Beland, Birczanin, Byronshock, C628, CardinalDan, Chris the speller, CsDix, D-Rock, DOHill, Danger, DavidLevinson, Defrenrokorit, Dhidalgo, Fylbecatulous, Gobonobo, Hroðulf, Hut 8.5, John of Reading, Jonathansammy, Jwhiting1234, Jyotsna.maddhukuri, Kevlar67, L0b0t, LilHelpa, Mari370, MelbourneStar, Morgan Riley, Mr.moyal, NedTugent, Norcalal, Nurg, Pgreenfinch, Radak, Rex the first, RiFraS, Richard.ohara, Schrauwers, Shakher59, Snori, Stew mss, Terrypin, Tomas e, Tractorboy60, Whoneedspants, Zlerman, 58 anonymous edits

Kerns *Source*: http://en.wikipedia.org/w/index.php?title=Kerns *Contributors*: Aliman5040, Alpöhi, Attilios, BillC, Charlie Tango Bravo, D6, Docu, Emerson7, InterwikiLinksRule, Jeppiz, Kokiri, Ksnow, Marek69, TicketMan, Tobyc75, 3 anonymous edits

Schattenhalb *Source*: http://en.wikipedia.org/w/index.php?title=Schattenhalb *Contributors*: Aliman5040, BillC, Chatfecter, Chris j wood, D6, Dbachmann, DidiWeidmann, Rigadoun, Starbois, Tobyc75, WereSpielChequers, 2 anonymous edits

Wolfenschiessen *Source*: http://en.wikipedia.org/w/index.php?title=Wolfenschiessen *Contributors*: Aliman5040, Bobblewik, CommonsDelinker, D6, DidiWeidmann, Kokiri, Ksnow, Marek69, Maximus Rex, Starbois, Tobyc75, 3 anonymous edits

Guttannen *Source*: http://en.wikipedia.org/w/index.php?title=Guttannen *Contributors*: Aliman5040, Andres, BillC, D6, Dbachmann, DidiWeidmann, Docu, MadGeographer, Rigadoun, SchuminWeb, Tobyc75, ZorkNika, 2 anonymous edits

Gadmen *Source*: http://en.wikipedia.org/w/index.php?title=Gadmen *Contributors*: Andres, BillC, Bonzon, D6, Dbachmann, DidiWeidmann, Docu, MadGeographer, Marek69, Rigadoun, Skew-t, Tobyc75, 1 anonymous edits

Grindelwald *Source*: http://en.wikipedia.org/w/index.php?title=Grindelwald *Contributors*: Ajho, Aliman5040, AllyUnion, Aranel, Attilios, BillC, Bryan Derksen, CMWF310, Chatfecter, Chochopk, Ckatz, Cometstyles, Cristo Vlahos, Cruccone, D6, David Shay, De728631, DidiWeidmann, Docu, Ekem, Elendil's Heir, Ericoides, Fairchildd, Glacier109, Halsteadk, Hapesoft, Hhk1989, Ikiwaner, Irmgard, Japanese Searobin, Jdhowens90, Jeppiz, Julien29, Jwissick, Kcarnold, Kenneth Stephen, Lajsikonik, MadGeographer, Malloryrdc, MichaelBillington, Naddy, Ncsakany, Nicolas1981, Nniuq25, Olivier, Osarius, PLCSales, Pp2007, Presidentman, Rama, RedWolf, Rich Farmbrough, Richard1966, Sabinpopa, Skew-t, Stan Shebs, Starbois, Steinninn, Ted

Wilkes, Thisisbossi, Thorwald, Tim bates, Timfpark, Tobyc75, TonyW, TracksOnWax, Tudorminstrel, Wilchett, Xezbeth, Zacharie Grossen, 51 anonymous edits

Engelberg *Source*: http://en.wikipedia.org/w/index.php?title=Engelberg *Contributors*: Accusativen hos Olsson, Aliman5040, Alpöhi, Andres, Attilios, BillC, Chithiraiyan, Chris j wood, ChrisGualtieri, D6, Derektamim, DidiWeidmann, Docu, Emerson7, Flayer, Gestumblindi, Ian Pitchford, Jeppiz, Junafani, Kam47625, Kevinalewis, KirkEN, Kokiri, Ksnow, Kuty, LeedsKing, Lostvalley, Lucifero4, Lzur, Marek69, Pjetter, Schigul95, Sfwparkes, Starbois, Strefli3, Swiss Ace, Tabletop, Tassedethe, Timberframe, Tobyc75, Williamb, Zacharie Grossen, Zigzig20s, 32 anonymous edits

Image Sources, Licenses and Contributors

File:Ausgang Aareschlucht.jpg *Source*: http://en.wikipedia.org/w/index.php?title=File:Ausgang_Aareschlucht.jpg *License*: unknown *Contributors*: Falk Lademann

file:Innertkirchen.JPG *Source*: http://en.wikipedia.org/w/index.php?title=File:Innertkirchen.JPG *License*: unknown *Contributors*: Aliman5040, Docu, Xöidi86

file:Switzerland location map.svg *Source*: http://en.wikipedia.org/w/index.php?title=File:Switzerland_location_map.svg *License*: unknown *Contributors*: Lukasb1992, Maoman, NordNordWest, Sting, Wikisoft*, 1 anonymous edits

File:Red pog.svg *Source*: http://en.wikipedia.org/w/index.php?title=File:Red_pog.svg *License*: unknown *Contributors*: Anomie

File:6140 - Innertkirchen - Hof-Sustenstrasse at Äppigenstrasse-Grimselstrasse.JPG *Source*: http://en.wikipedia.org/w/index.php?title=File:6140_-_Innertkirchen_-_Hof-Sustenstrasse_at_Äppigenstrasse-Grimselstrasse.JPG *License*: unknown *Contributors*: User:Thisisbossi

File:Karte Gemeinden der Schweiz 2012.png *Source*: http://en.wikipedia.org/w/index.php?title=File:Karte_Gemeinden_der_Schweiz_2012.png *License*: unknown *Contributors*: Tschubby

Image:Beatenberg-coat of arms.svg *Source*: http://en.wikipedia.org/w/index.php?title=File:Beatenberg-coat_of_arms.svg *License*: unknown *Contributors*: User:Aliman5040

Image:Bönigen-coat of arms.svg *Source*: http://en.wikipedia.org/w/index.php?title=File:Bönigen-coat_of_arms.svg *License*: unknown *Contributors*: User:Aliman5040

Image:Brienz-coat of arms.svg *Source*: http://en.wikipedia.org/w/index.php?title=File:Brienz-coat_of_arms.svg *License*: unknown *Contributors*: User:Aliman5040

Image:Brienzwiler-coat of arms.svg *Source*: http://en.wikipedia.org/w/index.php?title=File:Brienzwiler-coat_of_arms.svg *License*: unknown *Contributors*: User:Aliman5040

Image:Därligen-coat of arms.svg *Source*: http://en.wikipedia.org/w/index.php?title=File:Därligen-coat_of_arms.svg *License*: unknown *Contributors*: User:Aliman5040

Image:Gadmen-coat of arms.svg *Source*: http://en.wikipedia.org/w/index.php?title=File:Gadmen-coat_of_arms.svg *License*: unknown *Contributors*: User:Aliman5040

Image:Grindelwald-coat of arms.svg *Source*: http://en.wikipedia.org/w/index.php?title=File:Grindelwald-coat_of_arms.svg *License*: unknown *Contributors*: User:Aliman5040

Image:Gsteigwiler-coat of arms.svg *Source*: http://en.wikipedia.org/w/index.php?title=File:Gsteigwiler-coat_of_arms.svg *License*: unknown *Contributors*: User:Aliman5040

Image:Gündlischwand-coat of arms.svg *Source*: http://en.wikipedia.org/w/index.php?title=File:Gündlischwand-coat_of_arms.svg *License*: unknown *Contributors*: User:Aliman5040

Image:Guttannen-coat of arms.svg *Source*: http://en.wikipedia.org/w/index.php?title=File:Guttannen-coat_of_arms.svg *License*: unknown *Contributors*: User:Aliman5040

Image:Habkern-coat of arms.svg *Source*: http://en.wikipedia.org/w/index.php?title=File:Habkern-coat_of_arms.svg *License*: unknown *Contributors*: User:Aliman5040

Image:Hasliberg-coat of arms.svg *Source*: http://en.wikipedia.org/w/index.php?title=File:Hasliberg-coat_of_arms.svg *License*: unknown *Contributors*: User:Aliman5040

Image:Hofstetten bei Brienz-coat of arms.svg *Source*: http://en.wikipedia.org/w/index.php?title=File:Hofstetten_bei_Brienz-coat_of_arms.svg *License*: unknown *Contributors*: User:Aliman5040

Image:Innertkirchen-coat of arms.svg *Source*: http://en.wikipedia.org/w/index.php?title=File:Innertkirchen-coat_of_arms.svg *License*: unknown *Contributors*: User:Aliman5040

Image:Interlaken-coat of arms.svg *Source*: http://en.wikipedia.org/w/index.php?title=File:Interlaken-coat_of_arms.svg *License*: unknown *Contributors*: User:Aliman5040

Image:Iseltwald-coat of arms.svg *Source*: http://en.wikipedia.org/w/index.php?title=File:Iseltwald-coat_of_arms.svg *License*: unknown *Contributors*: User:Aliman5040

Image:Lauterbrunnen-coat of arms.svg *Source*: http://en.wikipedia.org/w/index.php?title=File:Lauterbrunnen-coat_of_arms.svg *License*: unknown *Contributors*: User:Aliman5040

Image:Leissigen-coat of arms.svg *Source*: http://en.wikipedia.org/w/index.php?title=File:Leissigen-coat_of_arms.svg *License*: unknown *Contributors*: User:Aliman5040

Image:Lütschental-coat of arms.svg *Source*: http://en.wikipedia.org/w/index.php?title=File:Lütschental-coat_of_arms.svg *License*: unknown *Contributors*: User:Aliman5040

Image:Matten bei Interlaken-coat of arms.svg *Source*: http://en.wikipedia.org/w/index.php?title=File:Matten_bei_Interlaken-coat_of_arms.svg *License*: unknown *Contributors*: User:Aliman5040

Image:Niederried bei Interlaken-coat of arms.svg *Source*: http://en.wikipedia.org/w/index.php?title=File:Niederried_bei_Interlaken-coat_of_arms.svg *License*: unknown *Contributors*: User:Aliman5040

Image:Oberried am Brienzersee-coat of arms.svg *Source*: http://en.wikipedia.org/w/index.php?title=File:Oberried_am_Brienzersee-coat_of_arms.svg *License*: unknown *Contributors*: User:Aliman5040

Image:Ringgenberg-coat of arms.svg *Source*: http://en.wikipedia.org/w/index.php?title=File:Ringgenberg-coat_of_arms.svg *License*: unknown *Contributors*: User:Aliman5040

Image:Saxeten-coat of arms.svg *Source*: http://en.wikipedia.org/w/index.php?title=File:Saxeten-coat_of_arms.svg *License*: unknown *Contributors*: User:Aliman5040

Image:Schattenhalb-coat of arms.svg *Source*: http://en.wikipedia.org/w/index.php?title=File:Schattenhalb-coat_of_arms.svg *License*: unknown *Contributors*: User:Aliman5040

Image:Schwanden bei Brienz-coat of arms.svg *Source*: http://en.wikipedia.org/w/index.php?title=File:Schwanden_bei_Brienz-coat_of_arms.svg *License*: unknown *Contributors*: User:Aliman5040

Image:Unterseen-coat of arms.svg *Source*: http://en.wikipedia.org/w/index.php?title=File:Unterseen-coat_of_arms.svg *License*: unknown *Contributors*: User:Aliman5040

Image:Wilderswil-coat of arms.svg *Source*: http://en.wikipedia.org/w/index.php?title=File:Wilderswil-coat_of_arms.svg *License*: unknown *Contributors*: User:Aliman5040

File:Suisse cantons.svg *Source*: http://en.wikipedia.org/w/index.php?title=File:Suisse_cantons.svg *License*: unknown *Contributors*: User:Pymouss44

File:Karte 13 Alte Orte.png *Source*: http://en.wikipedia.org/w/index.php?title=File:Karte_13_Alte_Orte.png *License*: unknown *Contributors*: User:Sidonius

Image:Wappen Zürich matt.svg *Source*: http://en.wikipedia.org/w/index.php?title=File:Wappen_Zürich_matt.svg *License*: unknown *Contributors*: Sa-se

Image:Wappen Bern matt.svg *Source*: http://en.wikipedia.org/w/index.php?title=File:Wappen_Bern_matt.svg *License*: unknown *Contributors*: User:Sa-se

Image:Wappen Luzern matt.svg *Source*: http://en.wikipedia.org/w/index.php?title=File:Wappen_Luzern_matt.svg *License*: unknown *Contributors*: Sa-se

Image:Wappen Uri matt.svg *Source*: http://en.wikipedia.org/w/index.php?title=File:Wappen_Uri_matt.svg *License*: unknown *Contributors*: Sa-se

Image:Wappen Obwalden matt.svg *Source*: http://en.wikipedia.org/w/index.php?title=File:Wappen_Obwalden_matt.svg *License*: unknown *Contributors*: Heraldiker Rudolf Münger

Image:Wappen Nidwalden matt.svg *Source*: http://en.wikipedia.org/w/index.php?title=File:Wappen_Nidwalden_matt.svg *License*: unknown *Contributors*: Sa-se

Image:Wappen Glarus matt.svg *Source*: http://en.wikipedia.org/w/index.php?title=File:Wappen_Glarus_matt.svg *License*: unknown *Contributors*: Sa-se

Image:Wappen Zug matt.svg *Source*: http://en.wikipedia.org/w/index.php?title=File:Wappen_Zug_matt.svg *License*: unknown *Contributors*: Sa-se

Image:Wappen Freiburg matt.svg *Source*: http://en.wikipedia.org/w/index.php?title=File:Wappen_Freiburg_matt.svg *License*: unknown *Contributors*: Sa-se

Image:Wappen Solothurn matt.svg *Source*: http://en.wikipedia.org/w/index.php?title=File:Wappen_Solothurn_matt.svg *License*: unknown *Contributors*: Sa-se

Image:Wappen Basel-Stadt matt.svg *Source*: http://en.wikipedia.org/w/index.php?title=File:Wappen_Basel-Stadt_matt.svg *License*: unknown *Contributors*: Sa-se

Image:Coat of arms of Kanton Basel-Landschaft.svg *Source*: http://en.wikipedia.org/w/index.php?title=File:Coat_of_arms_of_Kanton_Basel-Landschaft.svg *License*: unknown *Contributors*: Wappenentwurf: Otto Plattner diese Datei: Jwnabd

Image:Wappen Schaffhausen matt.svg *Source*: http://en.wikipedia.org/w/index.php?title=File:Wappen_Schaffhausen_matt.svg *License*: unknown *Contributors*: (previous version maybe User:Delta-9)

Image:Wappen Appenzell Ausserrhoden matt.svg *Source*: http://en.wikipedia.org/w/index.php?title=File:Wappen_Appenzell_Ausserrhoden_matt.svg *License*: unknown *Contributors*: Sa-se

Image:Wappen Appenzell Innerrhoden matt.svg *Source*: http://en.wikipedia.org/w/index.php?title=File:Wappen_Appenzell_Innerrhoden_matt.svg *License*: unknown *Contributors*: Sa-se

Image:Coat of arms of canton of St. Gallen.svg *Source*: http://en.wikipedia.org/w/index.php?title=File:Coat_of_arms_of_canton_of_St._Gallen.svg *License*: unknown *Contributors*: Wappenentwurf: unbekannt diese Datei: Jwnabd

Image:Wappen Graubünden matt.svg *Source*: http://en.wikipedia.org/w/index.php?title=File:Wappen_Graubünden_matt.svg *License*: unknown *Contributors*: Sa-se

Image:Wappen Aargau matt.svg *Source*: http://en.wikipedia.org/w/index.php?title=File:Wappen_Aargau_matt.svg *License*: unknown *Contributors*: Sa-se

Image:Wappen Thurgau matt.svg *Source*: http://en.wikipedia.org/w/index.php?title=File:Wappen_Thurgau_matt.svg *License*: unknown *Contributors*: Sa-se

Image:Wappen Tessin matt.svg *Source*: http://en.wikipedia.org/w/index.php?title=File:Wappen_Tessin_matt.svg *License*: unknown *Contributors*: Abaddon1337, Ricordisamoa, Sa-se, Skim, 1 anonymous edits

Image:Wappen Waadt matt.svg *Source*: http://en.wikipedia.org/w/index.php?title=File:Wappen_Waadt_matt.svg *License*: unknown *Contributors*: Sa-se

Image:Wappen Wallis matt.svg *Source*: http://en.wikipedia.org/w/index.php?title=File:Wappen_Wallis_matt.svg *License*: unknown *Contributors*: Sa-se

Image:Wappen Neuenburg matt.svg *Source*: http://en.wikipedia.org/w/index.php?title=File:Wappen_Neuenburg_matt.svg *License*: unknown *Contributors*: Sa-se

Image:Wappen Genf matt.svg *Source*: http://en.wikipedia.org/w/index.php?title=File:Wappen_Genf_matt.svg *License*: unknown *Contributors*: Sa-se

Image:Wappen Jura matt.svg *Source*: http://en.wikipedia.org/w/index.php?title=File:Wappen_Jura_matt.svg *License*: unknown *Contributors*: Sa-se

Image:Coat of Arms of Switzerland.svg *Source*: http://en.wikipedia.org/w/index.php?title=File:Coat_of_Arms_of_Switzerland.svg *License*: unknown *Contributors*: User:E Pluribus Anthony, User:Reisio, User:Wiki-vr, User:Zscout370

Image:Karikatur Teilung Basels.jpg *Source*: http://en.wikipedia.org/w/index.php?title=File:Karikatur_Teilung_Basels.jpg *License*: unknown *Contributors*: Alarob, Aliman5040, Infrogmation, Mapmarks, Mattes, Parpan05, Roland zh, Sidonius, Stefan Kühn

File:Brenodor.jpg *Source*: http://en.wikipedia.org/w/index.php?title=File:Brenodor.jpg *License*: unknown *Contributors*: User:Al-qamar

File:Engehalbinsel Römerbad1.jpg *Source*: http://en.wikipedia.org/w/index.php?title=File:Engehalbinsel_Römerbad1.jpg *License*: unknown *Contributors*: User:Ginkgo

File:Map Burgundian Kingdom 2 EN.png *Source*: http://en.wikipedia.org/w/index.php?title=File:Map_Burgundian_Kingdom_2_EN.png *License*: unknown *Contributors*: User:Sidonius

File:Schweiz Frühmia Adel.svg *Source*: http://en.wikipedia.org/w/index.php?title=File:Schweiz_Frühmia_Adel.svg *License*: unknown *Contributors*: User:Sidonius

File:Historische Karte CH 1385.png *Source*: http://en.wikipedia.org/w/index.php?title=File:Historische_Karte_CH_1385.png *License*: unknown *Contributors*: User:Sidonius

File:Historische Karte CH 1416.png *Source*: http://en.wikipedia.org/w/index.php?title=File:Historische_Karte_CH_1416.png *License*: unknown *Contributors*: User:Sidonius

File:Berner Aargau.png *Source*: http://en.wikipedia.org/w/index.php?title=File:Berner_Aargau.png *License*: unknown *Contributors*: User:Sidonius

File:Karte Stadtstaat Bern.svg *Source*: http://en.wikipedia.org/w/index.php?title=File:Karte_Stadtstaat_Bern.svg *License*: unknown *Contributors*: User:Sidonius

File:Belagerung Grandson.jpg *Source*: http://en.wikipedia.org/w/index.php?title=File:Belagerung_Grandson.jpg *License*: unknown *Contributors*: Johannes Stumpf, upload by

File:Karte Kanton Waadt Bezirke 2010.png *Source*: http://en.wikipedia.org/w/index.php?title=File:Karte_Kanton_Waadt_Bezirke_2010.png *License*: unknown *Contributors*: Tschubby

File:Karte Helvetik 3.png *Source*: http://en.wikipedia.org/w/index.php?title=File:Karte_Helvetik_3.png *License*: unknown *Contributors*: User:Sidonius

File:2009-01-11 View from Gurten near Berne in Winter 14.JPG *Source*: http://en.wikipedia.org/w/index.php?title=File:2009-01-11_View_from_Gurten_near_Berne_in_Winter_14.JPG *License*: unknown *Contributors*: User:Noebu

File:Lauterbrunnen Staubbach.jpg *Source*: http://en.wikipedia.org/w/index.php?title=File:Lauterbrunnen_Staubbach.jpg *License*: unknown *Contributors*: User:Parpan05

File:Joseph Anton Koch 003.jpg *Source*: http://en.wikipedia.org/w/index.php?title=File:Joseph_Anton_Koch_003.jpg *License*: unknown *Contributors*: Bukk, Concord, Docu, Emijrp, Goldfritha, Ixtzib, Rlbberlin, ZorkNika

File:Rathaus Bern - Session des Grossen Rates.jpg *Source*: http://en.wikipedia.org/w/index.php?title=File:Rathaus_Bern_-_Session_des_Grossen_Rates.jpg *License*: unknown *Contributors*: Andi Jacomet, WebFactory Staatskanzlei

File:Karte Kanton Bern Verwaltungskreise 2010.png *Source*: http://en.wikipedia.org/w/index.php?title=File:Karte_Kanton_Bern_Verwaltungskreise_2010.png *License*: unknown *Contributors*: Tschubby

File:bernriver.jpg *Source*: http://en.wikipedia.org/w/index.php?title=File:Bernriver.jpg *License*: unknown *Contributors*: Original uploader was Queenofthewilis at en.wikipedia

File:Emmentaler.jpg *Source*: http://en.wikipedia.org/w/index.php?title=File:Emmentaler.jpg *License*: unknown *Contributors*: Firefox13, IvanLanin, NeverDoING, Nordelch, Quadell, Ranveig, Tano4595, Túrelio, Wst, Zerohund, ZorkNika

Image:Elementy herbu.jpg *Source*: http://en.wikipedia.org/w/index.php?title=File:Elementy_herbu.jpg *License*: unknown *Contributors*: User:Qyd, User:Steifer

File:Hyghalmen Roll Late 1400s.jpg *Source*: http://en.wikipedia.org/w/index.php?title=File:Hyghalmen_Roll_Late_1400s.jpg *License*: unknown *Contributors*: AndreasPraefcke, Chatsam, Evadb, G.dallorto, Jaybear, Knorrepoes, LeoDavid, Massimop, Para, Rosenzweig, Skim, Technokrat, 1 anonymous edits

File:University of Plymouth Coat of Arms.png *Source*: http://en.wikipedia.org/w/index.php?title=File:University_of_Plymouth_Coat_of_Arms.png *License*: unknown *Contributors*: Garter, Clarenceux and Norroy and Ulster King of Arms.

File:Vaasa.vaakuna.svg *Source*: http://en.wikipedia.org/w/index.php?title=File:Vaasa.vaakuna.svg *License*: unknown *Contributors*: User:Care

File:Ottoman flag.svg *Source*: http://en.wikipedia.org/w/index.php?title=File:Ottoman_flag.svg *License*: unknown *Contributors*: AnonMoos, Cycn, David0811, FSII, Homo lupus, Quest for Truth, Ricordisamoa, TFCforever, TRAJAN 117, Молох, 1 anonymous edits

Image:Coats of arms of the Kingdom of Egypt and Sudan.png *Source*: http://en.wikipedia.org/w/index.php?title=File:Coats_of_arms_of_the_Kingdom_of_Egypt_and_Sudan.png *License*: unknown *Contributors*: User:Beao, User:CrimeCentral, User:SreeBot

Image:Coat of arms of United Arab Republic (Syria 1958-61, Egypt 1958-1971).svg *Source*: http://en.wikipedia.org/w/index.php?title=File:Coat_of_arms_of_United_Arab_Republic_(Syria_1958-61,_Egypt_1958-1971).svg *License*: unknown *Contributors*: User:AnonMoos, User:Kafka1, User:Omar86, User:R-41, User:Tonyjeff

File:Mitsubishi_logo.svg *Source*: http://en.wikipedia.org/w/index.php?title=File:Mitsubishi_logo.svg *License*: unknown *Contributors*: User:Monaneko

File:John paul 2 coa.svg *Source*: http://en.wikipedia.org/w/index.php?title=File:John_paul_2_coa.svg *License*: unknown *Contributors*: User:mAgul

File:PD-icon.svg *Source*: http://en.wikipedia.org/w/index.php?title=File:PD-icon.svg *License*: unknown *Contributors*: Alex.muller, Anomie, Anonymous Dissident, CBM, MBisanz, PBS, Quadell, Rocket000, Strangerer, Timotheus Canens, 1 anonymous edits

File:Guinea Dinguiraye farmers cooperative.jpg *Source*: http://en.wikipedia.org/w/index.php?title=File:Guinea_Dinguiraye_farmers_cooperative.jpg *License*: unknown *Contributors*: Béka, Dr. Blofeld, Mattes

File:JA-Hokkaido-Bldg-02.jpg *Source*: http://en.wikipedia.org/w/index.php?title=File:JA-Hokkaido-Bldg-02.jpg *License*: unknown *Contributors*: User:J o

File:Sugarcane weighing at sugarmill.jpg *Source*: http://en.wikipedia.org/w/index.php?title=File:Sugarcane_weighing_at_sugarmill.jpg *License*: unknown *Contributors*: User:Shakher59

File:Picswiss OW-27-30.jpg *Source*: http://en.wikipedia.org/w/index.php?title=File:Picswiss_OW-27-30.jpg *License*: unknown *Contributors*: Dake, Docu, Roland zh, Wiki4you

file:Kerns-coat of arms.svg *Source*: http://en.wikipedia.org/w/index.php?title=File:Kerns-coat_of_arms.svg *License*: unknown *Contributors*: User:Delta-9

File:Melchsee-Frutt Kuh.jpg *Source*: http://en.wikipedia.org/w/index.php?title=File:Melchsee-Frutt_Kuh.jpg *License*: unknown *Contributors*: User:Ikiwaner

File:Picswiss OW-27-33.jpg *Source*: http://en.wikipedia.org/w/index.php?title=File:Picswiss_OW-27-33.jpg *License*: unknown *Contributors*: Dake, Docu, Roland zh, Wiki4you

File:Hohe Brücke Obwalden 1893.jpg *Source*: http://en.wikipedia.org/w/index.php?title=File:Hohe_Brücke_Obwalden_1893.jpg *License*: unknown *Contributors*: Wiki4you

file:Schattenhalb-coat of arms.svg *Source*: http://en.wikipedia.org/w/index.php?title=File:Schattenhalb-coat_of_arms.svg *License*: unknown *Contributors*: User:Aliman5040

Image:Rosenlaui.jpg *Source*: http://en.wikipedia.org/w/index.php?title=File:Rosenlaui.jpg *License*: unknown *Contributors*: Docu, FSII, Ikiwaner, Voyager, ZorkNika

File:Trübsee.jpg *Source*: http://en.wikipedia.org/w/index.php?title=File:Trübsee.jpg *License*: unknown *Contributors*: User:Anniev

file:Wolfenschiessen-coat of arms.svg *Source*: http://en.wikipedia.org/w/index.php?title=File:Wolfenschiessen-coat_of_arms.svg *License*: unknown *Contributors*: User:Delta-9

File:Trübsee01.jpg *Source*: http://en.wikipedia.org/w/index.php?title=File:Trübsee01.jpg *License*: unknown *Contributors*: User:Cable1

File:Bannalpsee.jpg *Source*: http://en.wikipedia.org/w/index.php?title=File:Bannalpsee.jpg *License*: unknown *Contributors*: User:BraunW

File:OberrickenbachSeilbahn.jpg *Source*: http://en.wikipedia.org/w/index.php?title=File:OberrickenbachSeilbahn.jpg *License*: unknown *Contributors*: Mussklprozz at de.wikipedia

file:Guttannen-coat of arms.svg *Source*: http://en.wikipedia.org/w/index.php?title=File:Guttannen-coat_of_arms.svg *License*: unknown *Contributors*: User:Aliman5040

Image:Grimselpass.jpg *Source*: http://en.wikipedia.org/w/index.php?title=File:Grimselpass.jpg *License*: unknown *Contributors*: Cooper.ch, Docu, Ikiwaner, ZorkNika

Image:Grimselsee.jpg *Source*: http://en.wikipedia.org/w/index.php?title=File:Grimselsee.jpg *License*: unknown *Contributors*: Cooper.ch, Docu, FSII, Ikiwaner, ZorkNika

File:Gadmental ostwärts.jpg *Source*: http://en.wikipedia.org/w/index.php?title=File:Gadmental_ostwärts.jpg *License*: unknown *Contributors*: User:Parpan05

file:Gadmen-coat of arms.svg *Source*: http://en.wikipedia.org/w/index.php?title=File:Gadmen-coat_of_arms.svg *License*: unknown *Contributors*: User:Aliman5040

File:Sustenpass.jpg *Source*: http://en.wikipedia.org/w/index.php?title=File:Sustenpass.jpg *License*: unknown *Contributors*: Docu, Herzi Pinki, Noebu, Saperaud, 1 anonymous edits

File:Grindelwald View 02.jpg *Source*: http://en.wikipedia.org/w/index.php?title=File:Grindelwald_View_02.jpg *License*: unknown *Contributors*: User:Cristo Vlahos

file:Grindelwald-coat of arms.png *Source*: http://en.wikipedia.org/w/index.php?title=File:Grindelwald-coat_of_arms.png *License*: unknown *Contributors*: -

File:Grindelwald 1900.jpg *Source*: http://en.wikipedia.org/w/index.php?title=File:Grindelwald_1900.jpg *License*: unknown *Contributors*: DenghiùComm, Docu, Herzi Pinki, Howcheng, Jan Arkesteijn, MGA73, Roland zh

File:Grindelwald View.jpg *Source*: http://en.wikipedia.org/w/index.php?title=File:Grindelwald_View.jpg *License*: unknown *Contributors*: User:Cristo Vlahos

File:Alphorns Grindelwald.jpg *Source*: http://en.wikipedia.org/w/index.php?title=File:Alphorns_Grindelwald.jpg *License*: unknown *Contributors*: User:Cristo Vlahos

File:5438 - Grindelwald - New residence with garages.JPG *Source*: http://en.wikipedia.org/w/index.php?title=File:5438_-_Grindelwald_-_New_residence_with_garages.JPG *License*: unknown *Contributors*: User:Thisisbossi

File:Mh JB kleine scheidegg.jpeg *Source*: http://en.wikipedia.org/w/index.php?title=File:Mh_JB_kleine_scheidegg.jpeg *License*: unknown *Contributors*: Chris j wood, El., Gürbetaler, Juiced lemon, LosHawlos, ZorkNika

File:Kleine Scheidegg.jpg *Source*: http://en.wikipedia.org/w/index.php?title=File:Kleine_Scheidegg.jpg *License*: unknown *Contributors*: User:Grindel1

File:2012-07-26 Berner Oberland 5602.JPG *Source*: http://en.wikipedia.org/w/index.php?title=File:2012-07-26_Berner_Oberland_5602.JPG *License*: unknown *Contributors*: User:Noebu

File:WAB&BOB Grindelwald Station.jpg *Source*: http://en.wikipedia.org/w/index.php?title=File:WAB&BOB_Grindelwald_Station.jpg *License*: unknown *Contributors*: User:SalomonCeb

File:Engelberg.jpg *Source*: http://en.wikipedia.org/w/index.php?title=File:Engelberg.jpg *License*: unknown *Contributors*: User:Stan45

file:Engelberg-coat of arms.gif *Source*: http://en.wikipedia.org/w/index.php?title=File:Engelberg-coat_of_arms.gif *License*: unknown *Contributors*: User:Alpöhi

File:Hubert Sattler Blick auf Engelberg.jpg *Source*: http://en.wikipedia.org/w/index.php?title=File:Hubert_Sattler_Blick_auf_Engelberg.jpg *License*: unknown *Contributors*: AndreasPraefcke, Bukk, Docu, FA2010, Kurpfalzbilder.de, Mattes, Tohma

File:Engelberg Switzerland1.jpg *Source*: http://en.wikipedia.org/w/index.php?title=File:Engelberg_Switzerland1.jpg *License*: unknown *Contributors*: Self. Original uploader was Chithiraiyan at en.wikipedia

File:Engelberg-Titlis cableway wallpaper.JPG *Source*: http://en.wikipedia.org/w/index.php?title=File:Engelberg-Titlis_cableway_wallpaper.JPG *License*: unknown *Contributors*: Natan Flayer

File:GrossTitlisSchanzeNacht.jpg *Source*: http://en.wikipedia.org/w/index.php?title=File:GrossTitlisSchanzeNacht.jpg *License*: unknown *Contributors*: User:Mussklprozz

MIX
Papier aus verantwortungsvollen Quellen
Paper from responsible sources
FSC® C105338

Printed by Books on Demand GmbH, Norderstedt / Germany